BIOLOGICAL APOCALYPSE

BIOLOGICAL APOCALYPSE

Species Extinction in the Anthropocene

edited by

Pronami Bhattacharyya

PRIMUS BOOKS

PRIMUS BOOKS
An imprint of Ratna Sagar P. Ltd.
Virat Bhavan
Mukherjee Nagar Commercial Complex
Delhi 110 009

Offices at CHENNAI LUCKNOW
AGRA AHMEDABAD BENGALURU BHOPAL COIMBATORE DEHRADUN
GUWAHATI HYDERABAD JAIPUR JALANDHAR KANPUR KOCHI KOLKATA
MUMBAI PATNA RANCHI VARANASI

First published 2023

ISBN: 978-93-5687-865-5 (hardback)
ISBN: 978-93-5687-866-2 (POD)

Published by Primus Books

Laser typeset by Mithu Karmakar
mithu.karma@gmail.com

Contents

Illustrations

Maps

Figures

Plates

Tables

Foreword

In recent times a serious global concern about the loss of species and biodiversity, the unprecedented anthropogenic changes in the ecosystem, and the consequent direct or indirect implications of these developments on our planet has prevailed. There is a flow of interdisciplinary writings on the remarkable pace at which species loss has been occurring, relegating many into the categories of threatened, rare and extinct species in the Red Data Book. Much has been written about the collective existence and sustenance of all species in a cosmically networked global ecosystem. In her well-known work *The Silent Spring*, Rachel Carson revealed in the 1960s, how one single obnoxious chemical, Dichlorodiphenyltrichloroethane (DDT), could disrupt the life-supporting network of nature.

There are people, from both academia and the general public, who still think that these serious issues are nothing more than academic rhetoric with little practical significance. Even global leaders in some of the most powerful countries of the world, choose to dismiss current concerns about global warming, climate change and biodiversity loss as unfounded apprehensions. For many of them, economics is more important than ecology.

But how real and devastating these consequences could be, are being witnessed and experienced in the increasing frequency and intensity of natural calamities like droughts, super cyclones, floods, melting of glaciers, intense air pollution, etc., in different parts of the world today. The unprecedented scale of devastation, economic disruptions and loss of human lives due to such environmental catastrophes are hard evidences of this crisis.

Against this grim background, the present book, titled *Biological Apocalypse*, is like a breath of fresh air. Imaginatively conceived, this collection of essays focuses on interesting and important species—each either in the realm of extinction or in the critically threatened status. They provide graphic descriptions of the evolution, exploitation and finally, in some cases, extinction of the species. The book is a laudable attempt at creating a new level of awareness and sensitization, especially amongst future generations. The conceptualization of the book, the choice of the species and the grand narratives presented in these essays reflect the praiseworthy commitment of Pronami Bhattacharyya, the editor of the volume. With her background in English literature and her deep passion for the wild (especially for birds), Pronami has successfully infused a new sensibility towards wildlife protection through the various accounts included here.

It has been a very pleasant journey going through the chapters on various species covered in this small but beautiful book, and wish it all success in the world.

I look forward to more such useful works from Pronami in the future.

Kolkata RANJAN CHAKRABARTI

Acknowledgements

I EXTEND MY gratitude to the many people who helped me through the journey of ideating and curating this book.

I am deeply indebted to Ms Preeti Gill for placing her trust in me and directing my proposal towards my publisher. She has been a very strong and positive presence throughout the three years that the book has taken to shape up.

I am eternally grateful to Professor Ranjan Chakrabarti for consenting to write the Foreword to this book. It is a matter of great pride for me, as well as a remarkably humbling experience.

I must extend a special thanks to Mr Vikram Grewal for encouraging me to go ahead with this project when it was at its nascent stage. He has also been kind enough to readily provide me with lithograph of the Jerdon's Courser from his personal collection for the chapter on the species.

My unabridged appreciation goes to all the contributors of this book who promptly agreed to be a part of this project and contribute their respective chapters. Despite their occupied schedules, they have adhered to the timelines through the long exercise of three years. I thank them for being patient with me through this long stretch of time.

I would like to express my sincere gratitude to Dr Prasun Chatterjee, Deputy Head, Primus Books, for perceiving the substance of the book and the urgency of it. I also the had great pleasure of working with Ms Jyotika Mansata, Associate Editor at Primus Books, whose detailed and timely responses helped me noticeably consolidate the volume.

I'm deeply indebted to my family for encouraging me in all my endeavours, academic and otherwise. This book would not have been possible without the love, support, and blessings from them.

I would extend a special thanks to my husband, Mr Dipanjol Deka, for his continued and unfailing love, support, and understanding. Being the first reader of my book (in its draft form), I value his criticism of the work which has helped me calibrate it to a large extent.

Foremost, to the Almighty, I thank for all the love and blessings bestowed upon me, for being the guiding light throughout my academic journey and the grace of which has strongly made its presence felt in this book as well. It is this unwavering divine strength that has made this project see the light of the day.

While I am forever obliged to Mother Nature, I find myself terribly limited in expressing my affection and regard for her. Sadly, it is not her magnificence but her plight in the Anthropocene that prompted me to formulate this book. I hope this humble attempt goes a considerable way in meaningfully expressing my gratitude.

Lastly, I thank all those who crossed me at some point, in some form, in this journey of more than three years, and inspired me to keep going.

PRONAMI BHATTACHARYYA

Introduction

Biological Apocalypse in the Sixth Mass Extinction

> For humans to cause species to become extinct and to destroy the biological diversity of God's creation, for humans to degrade the integrity of the earth by causing changes in its climate, by stripping the earth of its natural forests, or destroying its wetlands, for humans to contaminate the earth's waters, its land, its air and its life with poisonous substances, these are sins.
>
> —E. Wilson, *The Future of Life*

> So long as the earth is able to maintain mountains, forests and trees
> Until then the human race and its progeny will be able to survive.
>
> —Durga Saptashati, 'Devi Kavacham', 54

After the novel coronavirus broke out in a remote province of Wuhan, China, in late December 2019, conspiracy theorists sprang into action to claim it was manufactured in a proximate lab. Almost two months into the virus attack, which had by then become a global pandemic, we knew that the virus may not have been engineered in a human laboratory, but humans nonetheless played a role, and a central one at that, in the pandemic.

At the beginning of the fourth month into the pandemic, I sat watching news on the television and condemning the ever-growing human carnival of foolishness (or hubris) in breaking all 'lockdown' rules, instrumental in caging the rampant spread of the virus. During one such fit of pique, I concocted the idea of this book. Amidst the market shelves full of successful books on wildlife and its crisis, I wanted an elementary collection of essays for young minds and adults alike. As torchbearers to a collective, impending future, I would not want our scions to feel cheated of seeing so many wonderful species of flora and fauna, which are currently being wantonly slaughtered by humankind. However, as I embarked on the journey of bringing out this book, the enthusiastic contributors, all adept in their respective fields, turned it into a more serious anthology, ranging from research studies to personal rendezvous with nature, to register global vanishings of wildlife.

Hunting, farming and the global movement of people from rural settlements to cities has led to colossal declines in biodiversity and amplified the risk of animal-borne dangerous viruses like Ebola, SARS, bird flu and now Covid-19. Pathogens quickly spill over from animals to humans and spread to far-flung areas. The essential cause of the Covid-19 pandemic is strongly suggested to be increased human contact with wildlife, amongst others, as these wild animals most likely carry the novel pathogens which spill over to human settlements, creating havoc.

In an influential study undertaken in 2008,[1] K. Jones and a team of researchers identified and listed 335 novel diseases that surfaced between 1960 and 2004. It was found out that at least 60 per cent of these were zootonic, derived from animals. The sudden spurt in such zootonic diseases is a direct upshot of the link between environmental damage and anthropocentric activities like logging, mining, road building through remote areas or forests, rapid urbanization and, of course, population growth. Human society is, thus, making incursions into the animal kingdom at an accelerated rate and coming into closer contact with animal species they may never have been near before. This entails the transmission of disease from wildlife to humans, at the hidden cost of human economic development.

A vital offshoot of the rhetoric of human 'progress' and a primary concern of our world and this book, is species extinction. As an aberration of the Anthropocene (named so because of the enormous and increasing human impact on the Earth), this rhetoric overlooks the right of other species to flourish alongside humanity. Rather, for centuries, it seems to have served to normalize extinctions. The human species, through its myopic lens, presumed these extinction events to be catastrophic for the species in question alone, and that they were not real cataclysms as long as the 'ecosystem functions' that benefitted humans remained unharmed. This is perhaps the biggest ethical lacuna that serves a catalyst for an 'all-species-extinction' in the future.

Trees have been the most influential preachers in my life. Growing up in the sprawling green tea estates of Assam (India), trees and the colour green hold a fortifying presence in my heart and soul. In one of the tea gardens, there was a Fragnipani tree outside my window. A short, plump tree with a couple of horizontal branches to climb on, it was my hideout on afternoons or lazy Sunday mornings. A common mynah was a constant visitor to my treehouse. After a while, the myna brought a friend with it, and we were three on the tree. I often acted as a host, treating my feathered guests with crumbled marigold biscuits. It was a perfect home to me, the two mynas, and even lot of ants. I also remember using the tree's beautiful yellow and white flowers to make beach ornaments for my school dance performance. When we were transferred to another garden, I almost went into a major melancholic phase because I had to leave this tree 'home' behind. I wondered who would feed the mynas. After almost three years I visited that bungalow on the occasion of a birthday party and went straight to the tree. It was gone; it had given way for the construction of an extended human habitation. That oxygen supplier, the home to many other species, was simply 'not there' like many other trees on this planet. As the Holocene ended and gave way to the Anthropocene, incalculable trees have been lost. Since our species began practicing agriculture around 12,000 years ago, we've cut down closely half of the world's estimated 5.80 trillion trees.[2]

The natural world is a web of elaborate connections, most of which go unnoticed by humans. But these connections are host to nature's finely maintained equilibrium for millions of years, and tinkering with any part of the chain can affect an entire ecosystem. While a small section of the human populace does realize this, the gravity of the situation does not hit us as it does on seeing a dead rhino smeared in blood and with its horns chopped off. For us to have an ontological understanding of how we, too, are an intricate part of this delicately maintained equilibrium, we need to realize that even the disappearance of a tiny, unknown insect in a remote corner of the earth can turn us into a 'thing of the past'.

Along with this, we must also comprehend that other species have value in themselves and a right to continued existence. Human beings should preserve them whether or not it is convenient or economically beneficial for themselves. We distinguish values into two fundamental types—extrinsic and intrinsic.[3] To place an extrinsic value on something is a reference solely to its usefulness for some human purpose. In conserving species of various kinds, we often rely on this value. But as myopic as our perspectives are, we forget that all species on earth, and not just Homo sapiens, have intrinsic as well as extrinsic values. Arguably, the invisible intrinsic values are more important than the visible extrinsic ones. Our planet's sorry state of affairs is based purely on the incompetence of humans to perceive this duality of values.

A healthy ecosystem is maintained by the biodiversity of its flora and fauna. Each time a species goes extinct, the ecosystem loses its balance and the world around us unravels a little. The penalty that the planet has to pay is profound, affecting all species, even us.

EXTINCTION: IT HAPPENS

Most of us are conversant with the idea of species extinction. The dodo and the dinosaurs are iconic examples of mass extinction (ME). Since its inception, Earth has witnessed myriad of species been rendered extinct by changing biological conditions. In fact, in estimation, almost 99.90 per cent of the species that have ever lived are extinct today.[4]

Not more than 200 years ago, the notion of species vanishing from the face of earth was considered unlikely; it was believed that species were fixed and that they did not evolve or die. Despite the fact that fossil hunting, a popular contemporary pastime, turned up many new forms of life, unknown to that time, species extinction was still not an accepted phenomenon. However, French naturalist, George Cuvier, challenged this orthodoxy in his famous 1813 work, *Essay on the Theory of the Earth*.[5] He put forward theories of almost biblical-scale deluges which wiped out entire species and replaced them with whole new sets. However, Cuvier's theory was challenged by many, especially British geologist Charles Lyell, who preferred gradual changes in Earth's history. By the mid-nineteenth century, 'Lyellian gradualism'[6] swayed popular notions and calamitous episodes backed by Cuvier was spun into a marginalized agenda.

An enthusiast of Lyell's gradualist school, Charles Darwin saw long-term progress as both the artefact of human exertion and nature's selection process. From his viewpoint, global diversity has been mostly constant through the geological time. Thus, evolution by natural selection occurs by accumulation of slight changes or improvements in the evolving species which posed as a threat to the existing, unchanged ones. Thus, he believed that the extinction of the 'less-favoured' species naturally followed. However, over 150 years into his 'survival of the fittest' theory we have arrived at a more nuanced view about the changes and constants of global biodiversity. While there have been many long intervals of minor changes, as Darwin envisaged, there have also been interludes of mass extinctions.

Over the 4.60 billion years of our planet's existence, there have been at least five major mass extinctions (ME),[7] including the Ordovician, the Devonian, Permian, Triassic, and Cretaceous mass extinctions. Each of these extinctions wiped out more than 75 per cent of the species present in that period. 'Mass extinction' is different from 'simple' extinction as, in ecology, the latter signifies the vanishing of an organism or group of taxa. In an ME, typically numerous species vanish simultaneously. There is a considerable upsurge in the volume of extinction taking into account more than one geographically widespread higher taxon during a comparatively short interim of geologic time. Currently, we are in the supposedly greatest period of ME. It is estimated that since 1500 over 900 plant and animal species have gone extinct,[8] a rate which is ever accelerating, and that it might take only 250–500 years from now for 75 per cent of existing species of birds, amphibians and mammals to go extinct.

MAGNITUDE OF EXTINCTION

All the species of animals that exist on the earth now will, at some point, become extinct. Some of these, through the process of evolution, may give rise to descendants—new forms to exploit altered environments—while others may simply vanish, leaving no line of descent. The consequence of the extinction of species depends largely on its role in the ecosystem. Predators, for instance, are the first ones to be affected by poaching or human economic greed. When a predator goes extinct, all of their prey are liberated from the predatory species, and this could have serious impacts on ecosystems. A number of grass-grazing animals like deer can destroy the vegetation cover of forests; a declining owl population can lead to an outbreak in the number of rodents; if rhinos vanish from an ecosystem, then it might bring adverse effects as their eating pattern helps in seed dispersal (they eat in one place, while defecate in some other place, which helps plants to disperse in the ecosystem). Thus, the loss of a species can affect the ecosystem in a number of ways. No matter how small or incomprehensible an organism is, it could impact our lives in far greater ways than we could comprehend. More than half of the prescriptions written worldwide contain names of medicines derived from some flora or fauna. Even penicillin, one of the first antibiotics used, was derived from a 'tiny' fungus. However, mindless exploitation to serve human need and greed has endangered several animal and plant species which have been traditionally used for their medicinal properties.

The dodo (*Raphus cucullatus*), has become a true icon of extinction, exemplifying one of the worst cases of human interference in an ecosystem. Since their discovery in the 1500s in Mauritius, dodos were often drawn as fat, sluggish, and defenseless birds. They were targets of popular lampoons. It was the perfect bird for Lewis Caroll's pictogram of off-beat animals in *Alice's Adventures in Wonderland*. The dodo that Alice encountered was 'faintly absurd . . . [and] spoke in words of many syllables'.[9] For centuries, thus, the dodo has been the subject of an 'exceptional amount of popular commentary, folklore and illustrations'.[10] The extinction of the dodo, thus, became a symbol of the extinction of inferior animals, emboldening the theory of the 'survival of the fittest'. This has also infamously transformed into referring to a person who lacks intelligence, is perplexed, or looks silly as a 'dodo'. However, studies have proven that anthropocentric activities set off the process of dodo extinction. Once humans arrived on the island of Mauritius, dodos were doomed. A number of 'mariner's accounts variously describe the capturing and eating of Dodos'.[11] The birds, being restricted in range, were easy to capture; thus, within a short time, Dutch colonists, sailors and visitors soon hunted down most of the dodo population. However, as it seems, hunting alone could not lead to the total annihilation of the species, as 'the population of Mauritius was never more than 50 people during the 17th century'.[12] As a species, dodos seemed to have a slow productive rate with susceptible eggs and chicks that are easily laid/found in the open and, thus, are vulnerable to predators. As such, they gave in to competition and predation by exotic (non-native) species such as rats, monkeys, goats, pigs and deer. Subsequently, in the latter half of the seventeenth century, the dodo became extinct for the same reason that animals, such as the passenger pigeon,[13] became extinct—human greed, negligence, and the exigencies of history. The dodo can also be considered one of the earliest examples of modern eco-vandalism: the impact of the Portuguese sailors on Mauritius not only wiped out the famous dodo, but further disrupted nature in unexpected ways. A popular Mauritian folklore holds that the seed of the Calvaria or Tambalacoque tree, having a thick endocarp (the innermost layer which surrounds the seed in a fruit), required passing through a dodo's gut in order to germinate.[14] However, this forest tree is now seriously threatened[15] and in 1941, R.E. Vaughan and P.O. Wiehe[16] suggested that the dodo must have been instrumental in the dispersal and germination of the seeds. S.A. Temple,[17] in 1977, further formalized the folk notion of dodo and the Calvaria mutualism. However, a segment of the research fraternity[18] has also put forward strong arguments against such an 'obligate mutualism'. But despite many serious doubts cast upon the mutualism of the dodo and the Calvaria, the popular notion that the seeds of the tree required treatment by a dodo's gizzard before their germination could take place, has thrived in ecological literature. Nevertheless, amidst divergent reports, the lesson that emerges is that the ecology thrives in a chain system and that the chain has to be preserved at all costs.

A 'wonder herb' like rosy periwinkle (*Catharanthus rosea*), which is native to the island of Madagascar, has been found to aid the body in healing malaria, or to support a body dealing with diabetes, cancer, and Hodgkin's lymphoma. Two very important cancer-fighting medicines are derived from the plant: vinblastine and vincristine. Vincristine has helped in increasing the chances of surviving childhood leukaemia from 10 per cent to 90 per cent, whereas vinblastine is used to treat Hodgkins' disease. Sadly, this plant is endangered in its own habitat in Madagascar due to heavy deforestation. On the commercial level, the worldwide sales of the medicines derived from this plant are worth over £75 million a year, but essentially none of this money trails back to Madagascar, which is one of the poorest countries in the world.[19]

Similar is the existential despair of the Houston toad, found exclusively in south-east Texas. It is threatened by habitat loss and modifications, and their young ones are killed in large numbers by the invasive fire ant. This toad is known for secreting chemicals through its skin, such as serotonin and alkaloids, to protect itself from predators which in turn proves extremely useful for medicines that treat heart and nervous disorders in humans. Its extinction, like various other species, will harm not just the ecosystem, but also result in a big setback for the medical world.[20]

Taking a huge leap back in time, Wonambi (*Wonambi naracoortensis*), a snake only found in Australia, became extinct almost 40,000 years ago.[21] In various cave sites in Australia, paleontologists have found the bones of a long-dead animal that belonged to an ancient group of snakes, all of which are now extinct. This group, the *Madtsoiidae*, survived for around 90 million years, from the middle of the Cretaceous to the Pleistocene ages. They were once found in Australia, South America, Africa, Madagascar, and Europe, but slowly died out until Australia was their last refuge—Wonambi was one of the last of their number. Wonambi probably loitered around watering holes and other places that attracted its prey. Like the living giant snakes, Wonambi was probably non-venomous, instead relying on ambush tactics and its muscular body to catch and suffocate prey using constriction. Wonambi was a primitive snake, and it lacked the highly flexible skull of modern snakes; therefore, it was probably limited to smaller prey such as the many species of smaller wallaby that still inhabit Australia. They seem to have died out with the rest of the Australian megafauna around 40,000 years ago, but as new evidence comes to light, this date may change significantly. Humans may have known these snakes, and it is possible that human activities, such as bushfires, led to their demise. Australia, like the rest of the world, has been through some massive climatic changes in the past 2 million years or so, and perhaps the demise of these snakes coincided with the disappearance of the lush vegetation that once shrouded the Australian continent, leaving the arid landscape we know today. Water holes and other habitats favored by Wonambi disappeared, and its prey grew increasingly difficult to find. Confronted by this changing world and the pressure of human hunting, the Wonambi and the other primitive snakes eventually disappeared.[22]

Today, we have a series of such well-documented and well-acknowledged mega-crises, presenting a much-clearer picture of the evolutionary history of this planet. The world knows that prosperous dominant groups, like the dinosaurs, can be eliminated and replaced by obscure beings like the mammals. Though the catch phrase, 'survival of the fittest', has become a fixture of our culture, and is still considered a valid process of species extinction, several anthropocentric activities have added newer concerns/reasons, with habitat loss being the greatest of all.

While extinctions occur naturally, human modifications to the planet's ecosystem in the last millennia have critically accelerated the rate of extinction, which is much greater than the natural rate seen in the past. This may lead us to a point of great human-driven (anthropocentric) mass extinction, which is a complete anomaly, unlike the natural mass extinctions of historic the past.

EXTINCTION: ITS CAUSES

It is seen that traditional explanations normally do not take into account the genetic or cellular mechanisms of the mass die-offs; rather, they rely on Darwinian mechanisms underscoring the twin theories of 'survival of the fittest' and 'natural selection'. Nevertheless, be it natural selection, luck, or the apoptosis, which is normal in the evolutionary process, there is a general consensus on the following causes of extinction of species:[23] global warming, fluctuations in sea level, major glaciation, global anoxia, volcanic eruptions, plate tectonics, asteroid, comets and meteor impacts, gamma rays, and disease. These factors, often in combination with others, are considered to be the major causes of almost all the five major mass extinctions hitherto discussed. Most widely known, perhaps, is the K-T extinction which rendered the dinosaurs extinct. It is believed to have been triggered by a major environmental alteration that occurred after an asteroid impacted the Earth in the Cretaceous period almost 66 million years ago. Massive volumes of dust were released into the atmosphere, obstructing the sun, and ultimately leading to global cooling. Hard or slushy, experts variously believe that the earth has been turned into a giant snowball at least thrice so far.

THE BIG FIVE

The Ordovician period (485 to 444 million years ago) was brief glacial interval of histrionic changes for life on Earth. Species diversity thrived over a time span of 30 million years, but towards the end of the period, the first known mass extinction occurred. Massive glaciation led to locking up of huge amounts of water in the immense ice cap that covered large parts of the South Polar Region. These freshly uplifted rocks, due to large-scale weathering, accumulated carbon dioxide from the atmosphere which radically cooled the planet, turning it into the first snowball.[24]

This resulted in the sea levels plummeting by hundreds of feet. Almost 85 per cent of marine species, such as corals, shelled brachiopods, eel-like creatures called conodonts, etc., perished. This as the second-worst ME known to science. Creatures living in shallow waters must have received a dramatic blow on account of the drastic habitat cooling.

The Late Devonian extinction (383 to 359 million years ago) eradicated about 75 per cent of all species on Earth over a span of around 20 million years. It has been difficult to nail down the cause for the extinction pulses of this epoch, but volcanism is a potential trigger. In one of the pulses, ocean oxygen levels plunged precipitously, resulting in serious blows to conodonts and gonianites (ancient, shelled relatives of squid and octopuses). Many reef-building creatures were also eradicated, comprising a primary group of sea sponges called the stromatoporoids.

The Permian-Triassic extinction (252 million years ago) or the 'Great Dying' is a cataclysmic (and single-worst) event on earth that wiped out about 96 per cent of all marine species and about three of every four species on land. Even the world's forest ecosystems were completely wiped out and took almost 10 million years to recover, while the marine ecosystems took almost 4–8 million years to convalesce. The extinction's single biggest cause was a massive volcanic composite that erupted over more than 720,000 cu. mi. of lava across what is now known as Siberia. It resulted in very last ounce of fossil fuel on Earth being simultaneously dug up and burned, which might have led to a emission of large amounts of greenhouse gases such as methane. The consequential global warming was unequivocally hellish. In the succeeding million years after the cataclysm, seawater and soil temperatures jetted between 25 to 34° Fahrenheit. By 250.50 million years ago, surface temperatures of the sea, at the Equator, elevated to almost 104° Fahrenheit, turning it into a hot tub and leading to the vanishing of all fishes near the Equator. Various climate models propose that at the time, the oceans lost an approximate 76 per cent of their oxygen roster. These models also suggest that the unprecedented warming and the eventual oxygen loss account for a majority of species going extinct.

The Triassic-Jurassic extinction (201 million years ago) hit the earth with a sudden loss of up to 80 per cent of all marine and land species. However, a massive and rapid diversification of species had taken place after the 'Great Dying' which went on for millions of years. Various reef-building creatures began to flourish, and lush vegetation blanketed the earth. By the Triassic age, this was a perfect setting for a group of reptiles called the archosaurs to arrive on the scene: the precursors of birds, crocodilians, pterosaurs, and the non-avian dinosaurs. Towards the end of this epoch, the Earth warmed between 5 and 11° Fahrenheit, resultant in a quadrupling of atmospheric carbon dioxide levels. The marine creatures could no longer build their shells from calcium carbonate. On land, the dominant vertebrates, the crocodilians, which were bigger and far more varied than they are today, died out. In their wake, the earliest small and nimble dinosaurs on the ecological periphery briskly diversified.

The Cretaceous-Paleogene extinction (66 million years ago) is the most recent mass extinction and the only one conclusively associated with a major asteroid impact. Almost 76 per cent of all species on the planet, including all non-avian dinosaurs, went extinct. An 'epic' asteroid crashed into the sea just offshore of Mexico's Yucatán Peninsula, triggering a nightmarish sequence of events. The colossal impact left a crater of more than 120 mi. wide, tossing enormous volumes of dust, debris, and sulfur into the atmosphere. This resulted in a severe global cooling leading to an almost-nuclear winter. A huge tsunami swelled outward from the impact, while huge wildfires kindled chunks of land within 900 mi. of the impact. The ecosystems that sustained non-avian dinosaurs began to collapse overnight.

Mega-volcanoes are also major suspects in mass extinctions. Global warming caused by massive volcanic eruptions at the Deccan Flats in India may have aggravated the Cretaceous event. Some scientists even contend that some of the Deccan Flat eruptions could have been initiated by the impact.[25]

The Sixth Extinction

> Biodiversity—the diversity within species, between species and of ecosystems – is declining faster than at any time in human history.
>
> —*Global Assessment Report on Biodiversity and Ecosystem Services of the Intergovernmental Science-Policy Platform on Biodiversity and Ecosystem Services*

The Anthropocene (10,000 years ago to current times) defines earth's most recent geologic time period, which is characterized by major alterations in the physical environment owing to human or anthropogenic influences. The term 'Anthropocene' combines the root 'anthropo', meaning human, and the suffix 'cene' is a standard usage in geologic time as a referral to epochs.

Studies show overwhelming global evidence that atmospheric, hydrologic, geologic, biospheric and all other earth system processes are being hugely impacted by human activities.[26] Humans have very clearly remodelled earth since the last ice age ended, which is marked as a geologic time switch to the Holocene. However, at a conference in 2000,[27] Nobel Prize-winning atmospheric chemist Paul Crutzen claimed that we are precisely in the Anthropocene. Although the two terms, Holocene and Anthropocene, are often synonymously used, it is essential to call for some serious deliberation here.

In the last 12,000 years of a relatively stable climate, since the last ice age, human civilization has developed and we now have various narratives defining and redefining the emergence of our race. From the prehistoric period, the human role in nature—as progeny, curator, companion, gardener, or slayer—has grown with and within our ecology. However, overpowering evidence now confirms that humans are altering the Earth in unprecedented ways. The conspicuous hastening in our 'progress' since the mid-twentieth century—leaving a negative footprint of excess carbon dioxide emissions, excessive sea level rise, the mass extinction of species at a global level, and the alteration of land by deforestation for countless 'developmental' activities—mark the end of that slice of geological time which we have known as the 'Holocene'. Thus, it has been variously argued and widely accepted that the Holocene has given way to the Anthropocene.

Humans need space and resources, as do all other living beings. However, Kevin Mckee, in *Sparing Nature*,[28] explains that the growth rate of human population is far greater and imbalanced, and hence, unnatural, when matched to the growth of any other species in nature. This unnatural growth entails the construction of new homes to house this population, new office buildings, new malls, as well as new roads and new parking lots to ease our way around during our day-to-day activities. All this takes up space, covering lands what was once rich farmlands or forests. While Mckee is talking about his native land, Ohio, a developed place (nation), even in developing countries like India, deforestation poses a major threat to the ecology, causing an intolerable imbalance in nature. In a study undertaken in 2018,[29] it was found that about 75 per cent of Arunachal Pradesh is covered by a dense forest canopy with a varied type of covers including tropical, sub-tropical, temperate and alpine. This indicates towards the state's highly diverse climate. Sadly, there is an ever-growing pressure on forest resources owing to a consistent rise in the population, subsequent 'development' activities, huge number of wood-based industries and practically unsustainable land-use practices. This has also resulted in a sharp decline in the forest cover of the state, and forests, as we know, are home to a large number of species of wildlife. Several disturbances were observed during the field survey in the forests of Arunachal Pradesh from 2007 to 2010. As recently as 2017–18, there have been further 'developments' here. Being a birder, I frequent remote parts of north-east India along with my husband. Over the last three to four years, we have come across a sudden introduction of and spurt in the use of the electric saw in Arunachal Pradesh. We have encountered people felling trees and clearing spaces in

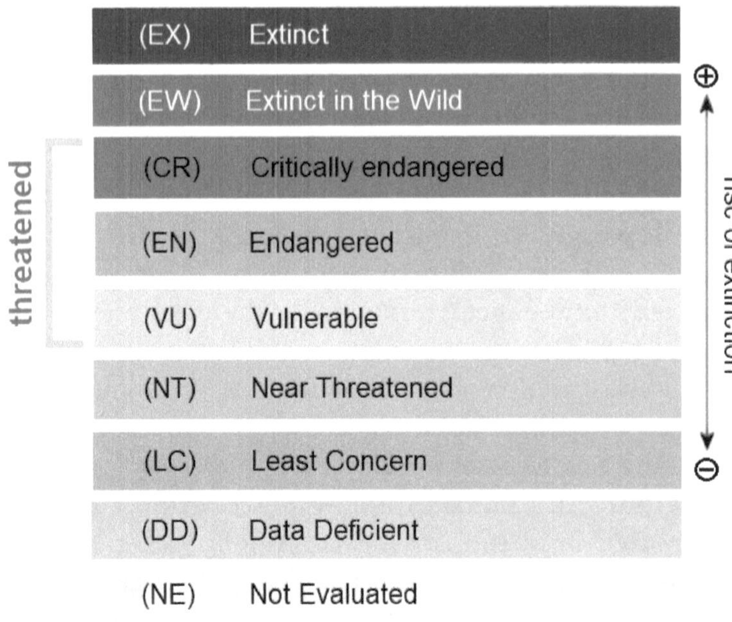

FIGURE I.1. IUCN Red List Structure
Source: Wikimedia Commons

several places along the route from Dirang 'Basti' to Mandala top. As a result, we have observed a marked decline in the availability of various rare avian species, like the laughing thrush or tragopan, or the mammal native to the state, the Arunachal macaque, which is rendered 'endangered' according to the International Union for Conservation of Nature (IUCN).

Arunachal Pradesh is also a potential habitat to the red panda, also known as the lesser panda or red bear-cat. However, habitat destruction and hunting has led it to be listed in Schedule I of the Wildlife (Protection) Act while being categorized as an endangered species in the IUCN Red List. To put it simply, this beautiful valley is losing its charm at an accelerated pace.

Anthropocentric actions threaten more species with worldwide extinction now more than ever before. An average of around 25 per cent of species in (catalogued) animal and plant groups are vulnerable, projecting that around 1 million species already face total annihilation, and many within decades, unless necessary action is taken to arrest and reduce the intensity of catalysts of biodiversity loss.[30]

Some people perceive the loss of farmland and get worried about the loss of space for human food to be cultivated. Lesser farmland and growing industries are definitely a concern for declining food production. While this may seem to be a legitimate concern for now, what is more alarming is the that the loss of agricultural land is the death knell of what existed there before it was cleared for cultivation of food crops and livestock—the forest cover. Forests are vast ecosystems in themselves, being abodes to an amazing variety of creatures, all interrelated. The extinction of one of those leads to a general imbalance the entire ecosystem. Thus, as in the words of Mckee[31] 'nature's land became our land'[32] and we continue subsuming nature as an end to serve our ever-growing needs. The biggest concern is the consequent demise of this biodiversity through extinctions, great and small.

Taking into account the past five mass extinctions, the current crisis illustrates tangible consequential losses on all fronts: ecological, spiritual and cultural. A 2009 study on 'Causes and consequences of Species Extinction' presents a grim picture of how rapid the human-induced global environment changes are:

> The five largest mass die-offs in which 50–95% of species were eliminated occurred during the Ordovician (490–443 million years ago [mya]), Devonian (417–354 mya), Permian (299–250 mya), Triassic (251–200 mya), and Cretaceous (146–64 mya) periods. Most recently, human actions especially over the past two centuries have precipitated a global extinction crisis or the 'sixth great extinction wave' comparable to the previous five. Increasing human populations over the last 50,000 years or so have left measurable negative footprints on biodiversity.[33]

The current extinction wave seems to lead towards an even more ominous future than those caused by the Big Five. Does an age of humans, thus, mean the end of nature? Yuval Noah Harari once quoted, 'When the first humans reached Australia about 45,000 years ago, they quickly drove to extinction 90% of its large animals. This was the first significant impact that Homo sapiens had on the planet's ecosystem. It was not the last.'[34] And yet, we march on. Since humans started tilling the land around 12,000 years ago, nearly half of the world's estimated 5.80 trillion trees have been cleared.[35] 'Even a single, isolated tree in an open area can act as a biodiversity "magnet," attracting and providing resources for many animals and plants. Therefore, losing even individual trees can severely impact biodiversity locally.'[36]

So, what could be the possible signal for future geological records to define the Anthropocene? Looking back, for example, the Cretaceous epoch claimed a 'golden spike' in sediments of metal iridium around the world, which was disseminated from the meteorite impact, leading to the historic extinction of the dinosaurs. For us, we are spoiled for choice, with our ingenuity in creating the negative footprint we are leaving behind. Among the golden spike for us could be the radioactive elements in the stratosphere and on earth due to nuclear tests, negative carbon footprint, aluminium and concrete elements, plastic pollution, and high levels of nitrogen and phosphate in soils, derivatives of artificial fertilizers, etc. The list is unbounded.

What we leave for the future is a clear Frankenstein being manufactured in the laboratories of human civilization. Our future is clearly leading to a mass extinction of species. The 'biological annihilation'[37] of wildlife, accelerated in past few decades, is clear indication that the Earth is on its way to a sixth mass extinction, which could be more severe than the previous five.

Scientists have undertaken studies of both rare and common species and found that millions of them have gone extinct without us even noticing. The single-most pressing cause for the crisis is human overpopulation and the consequent overconsumption. This has also paved the way to a severe overlapping of human and animal (wildlife) territories. Factors like habitat destruction, invasion by alien species, overhunting, toxic pollution, and climate change are obvious other factors, but the definitive cause of all of these factors is 'human overpopulation' and its 'continued growth'. The crisis is a direct threat not only to the animal kingdom, but to the very survival of the human civilization, leaving us with just a short time window in which to act.

This entails an important understanding of the delicate, almost invisible chain that exists in the natural world. Even small interferences in nature can have colossal consequences, and we must avoid anything that we have no demanding reason to touch.

NATURE'S CHAIN

Nature is a friend, a revered mother, a father to be obeyed and a beloved child to be nurtured[38]—such is the deep sense of communion between human and nature, as is evident in the Upanishads, the Vedas, and the Puranas (the sacred Indian texts). This natural relation is sacred because man depends on it and this sanctifies everything in nature, including its frightening aspects, such as landslides, storms, and earthquakes. All natural phenomena are seen as the manifestations or expressions of the gods. The basis of Hinduism, Buddhist and Jainism as a way of life capitalizes on the concept of dharma or righteousness which integrates duty, justice and cosmic law. Sanatana Dharma of the Hindu of way of life propagates supporting the whole universe. A 'dharmic' person indulges in the well-being of all living creatures. Thus, unlike the concept of Anthropocentricism, ancient knowledge celebrates human–nature relationship. Nandita Krishna, in her book *Hinduism and Nature*,[39] offers a remarkable reading of the integral correlation between mankind and nature. The necessity to protect and conserve our biodiversity is epitomized in the representation of the habitat of Lord Shiva, where he resides with his consort

Parvati and his two sons Karttikeya and Ganesha. His habitat, the mighty Mount Kailas, with snowy peaks, represents the cosmic heavens. She elaborates:

> The crescent moon on his forehead denotes tranquillity; the constant stream of Ganga's water from the lock of hair on his head indicates the purity and importance of water; Nandi, his bull mount, represents the world of animals; serpents signify the presence of toxicity in nature; the lion used by his consort Parvati represents wildlife; the peacock, the mount of Karttikeya, represents the avian species; and the mouse, the mount of Ganesha, represents small underground animals. Different types of animals and birds inhabit the holy abode of Lord Shiva. Another significant aspect is the harmonious relationship between natural enemies. In Lord Shiva's household, various natural enemies live in harmony with each other. The carnivorous lion's food is the vegetarian bull, the peacock is the enemy of the serpent and the mouse is the serpent's food; nevertheless, all live together.[40]

Similarly, one can also see a beautiful relationship between nature and mankind in Greek mythology and the organic way in which the two interact. However, on the flipside, Greek mythology also depicted the intermittent not-so-happy pairing of nature and mankind. Apart from nature being symbolic of life-giving properties, a number of myths stress the need for humans to treat nature with respect and humility. For the ancient Greeks, Earth was addressed as 'Gaia',[41] the Earth goddess (as in 'Prithvi' or Mother Earth in Hinduism), who creates and sustains all of nature. Prometheus created humans out of clay, and enabled them to thrive by giving them 'fire' stolen from the gods. Gaia and Prometheus, thus, play important roles in creating the Anthropocene.

The Biblical story of Genesis shows how, despite human's privileged central role in God's creation, they are nevertheless, following the Fall, are committed to toil in cultivating the earth. Genesis 2.15 says: 'The LORD God took the man and put him in the garden of Eden to till it and keep (preserve) it.'[42]

The art of listening to the speech of the animals has been kept alive in folklores of various cultures. In fact, in the mythologies of the world, the ability to comprehend the language of the birds was regarded as a divine boon. For instance, Odin, the Norse god, supervised the world with the help of two wise ravens, Huginn and Muninn.[43] Similarly, in tales from Wales, Greece, India, and greater Russia, one comes across an unlikely hero who is granted the gift of understanding birds by magic; while in Sweden, Sigurd slays a dragon, tastes its blood and hears the song of nature (i.e. language of birds).[44] King Solomon's widely revered wisdom, like that of Odin, was accredited to birds,[45] as were the fables of Aesop. However, for modern urban humans, these are generally assumed to be fabled tales and stories for children.

An early Indian literary work of the classical Sanskrit writer Kalidasa (fifth century), *Abhijana Shakunthalam*,[46] stands first for the sublime relationship between human and nature. To him, both are one organic entity and the order is first nature, then man. A detailed study of the work envisages a deep eco-consciousness, a trend which cuts across the literature of the world. The English bard, William Wordsworth, is greatly lauded as a nature pastor. He perceived an ethereal vigor in all the elements of nature. His poem, 'Tintern Abbey'[47] has nature or the environment as its prime theme. It is a landscape poem which edifies and celebrates the mysterious silence of nature. Popularly known as a pantheist, the eighteenth-century poet seems to have realized the dangers of annihilating nature and warned mankind against it through his subtle, yet profound writings.

Nature's Clockwork-like Mechanism

This cursory tour around the world through myths and literature brings us to a point where we can successfully find that the prototypes for living well together on the planet, based on reverence, reciprocity, responsibility, relationship, and respect, are manifest in every traditional and indigenous way of life. Like a giant clockwork-like mechanism, everything in nature is neatly arranged and interconnected. All animals and plants are part of an elusive equilibrium, and every entity has its purpose and role in its ecosystem.

Peter Wohlleben, in his book *The Secret Network of* Nature,[48] talks about the wolf, for instance. Under the general order Carnivora, there is a suborder called Caniformia, which comprises the family Canidae and the subfamily Caninae, which embraces the category Canis, and within that category we finally come to the species wolf. A long chain indeed! As predators, wolves keep a check on the number of plant eaters so that deer populations, for instance, do not multiply too swiftly. Thus, if we get rid of the wolves, we not only get rid of the biggest of enemies of cattle ranchers, perhaps, but also bring about a general imbalance in the nature's balance of work so much so that either a species of some birds may get eradicated from that ecosystem or a ravine or river either dry out or change its course. Phew!

The Curious Case of Yellowstone Wolves

For centuries, the wolf has been the source of long-standing myths and legends across the world, making its way into several successful books and movies. In a similar fashion, it has become almost a legendary specimen of disruption in nature's chain and its consequent affects. On such example can be seen in the history of America's first national park, Yellowstone National Park.[49] It is as if wolves and Yellowstone have become synonymous.

In the beginning of the nineteenth century, people began to systematically eliminate wolves in the park, principally in response to pressure from ranchers in the neighbouring region, who were anxious about their grazing livestock. The last pack of wolves was wiped out in 1926, even though individual wolves continued to be sporadically spotted until the 1930s, post which the predators were completely eliminated. Once the wolves vanished, the elk population began to grow, so much so that over the next few years, the condition of Yellowstone National Park deteriorated radically. Huge areas of the park were stripped bare by the elks, who were voracious grazers. The succulent grass by the river vanished, along with all the growing saplings. This resulted in a desolate landscape that didn't offer enough food even for birds, and the number of species vanished in significant numbers. Beavers, the worst affected, lost their food source and the lumber to build their dams. Also, since all the fresh deciduous trees beside the water were being eaten up by the hungry elk population, the beavers had nothing to gnaw on. Riverbanks tuned into wastelands deprived of any vegetation to protect the ground. This resulted in seasonal flooding and the eventual rapid increase in soil erosion. Thus, the meandering rivers cut across the landscape with their ever-increasing serpentine paths. To manage this situation, a part of the elk population was captured and shipped to other locations, while park authorities and hunters played their part in culling and killing the population. As a result, by the 1960s the elk numbers reached a historic low, causing outrage amongst conservationists and hunters alike. The culling and hunting stopped. As a result, in twenty years' time, the elk population once again reached unprecedented heights and overgrazing resurged.

This is a clear indication that what missed in the 1940s and 1950s was a sense of what we now call ecology. This sorry state of affairs continued for decades, and finally in 1995, wolves caught in Canada were released in Yellowstone to revive its lost ecological balance. However, what happens when a species, after its complete removal from an ecosystem, is reintroduced abruptly after a long interval? The wolves had an abundance of elk to feed on. The wolves' predation led to the dwindling of the elk population for the second time. As a result, the lower elk population provided an opportunity for trees to grow again. But what ensued was not very positive. With the decline in the elk population, the wolf packs faced difficulty in finding prey, leading them to either leave the park or starve. Along with this, a difference was also marked in the elk's behavior. The herbivores that had the entire park to graze carefree for decades were now unexpectedly attacked and devoured by the predators. Triggered by fear, they avoided open spaces and riverbanks, keeping mostly to covert areas. As a result, the riverbanks began to revive, and species such as birds, beavers, mice and bears returned. With the reintroduction of just a small population of wolves, the ecological scenario of the whole park transformed.

Today, we have several reports that claim that Yellowstone hasn't reclaimed into its original, pristine ecological state. Many claim that with the altered conditions in the intervening period, the reintroduction of the predators has not restored the ecosystem completely. There are signs of both promise and limitations. However, from the point of view of park management and biologists, the Wolf Recovery Project is considered a major success, in fact, it is seen as one of the most successful wildlife projects in US history.

Invasive Species

Introducing alien/invasive/non-native species can make things go awry in a new ecosystem, since every ecosystem is delicately intertwined and complete in itself. Burmese pythons, for instance, are the top predators in the Everglades National Park, Florida. As such, they have almost annihilated local mammal and bird populations. Skilled in consuming deer and even species as big as the alligators, these creatures eat practically any animal they encounter.[50] In the Western Ghats in India, for example, lantana (*Lantana camara*), a Central and South America shrub introduced as an ornamental flora in the 1800s resulted in a marked reduction in the abundances of non-timber forest products (NTFPs).[51] Lantana distresses revival of other forest plants that wild herbivores depend upon. It has also invaded over 40 per cent of India's tiger habitat and 300,000 sq. km. of Indian forests in general.[52] Australia, on the other hand, is experiencing one of the worst extinction events of mammals it has ever seen; this is largely due to an invasion of European red foxes, purposefully introduced for recreational hunting in 1855, and feral cats to control the rat menace.[53]

The diversity of ant species in an ecosystem perform important ecological roles, from scattering seeds to breaking down wood and returning essential nutrients to the soil, they are nature's best gardener. However, the constructive gardener can also prove disastrous in the role of an invasive species. For instance, a 2001 University of California study[54] shows how the tiny black Argentine ant can turn into a household pest. They can also disrupt natural ecosystems by replacing native ants. This study was instrumental in showing that when keystone (beneficial) species are killed by an invader, the negative effects can echo through the ecosystem. Similarly, yellow crazy ants can also prove to be a menace. Invasive ants can disturb native ant–plant mutualisms, such as pollination and seed dispersal, and change species composition. The yellow crazy ant is registered by the International Union for the Conservation of Nature's Invasive Species Specialist Group (IUCN-ISSG) among a hundred specimens of some of the 'world's worst invasive species'.[55] It has been reported as a pest in East Africa, Papua New Guinea, Malaysia (Borneo), Indonesia, Japan, Réunion, Mauritius, Rodrigues, Agaléga and India.[56]

Knowledge of physiognomies is helpful in screening potential invaders and in elaborating strategies to limit their success. A species' traits determine its success or failure in the transition between the different stages of the invasion process, and only particular trait combinations are assumed to make a species invasive.[57] Invasive species are primarily spread by anthropocentric activities like people's mobility and goods transportation. Thus, an amphibian (like the cane toad), plant, fungus, bacteria, fish, insect, or even an organism's seeds or eggs travel around the world very quickly via humancentric modes and often carry uninvited species with them. Ships can carry aquatic organisms. Pests can get into wood, shipping palettes, and crates that are shipped all over the world. Some ornamental plants can grow into the wild and turn invasive. And some invasive species are intentionally or accidentally released as pets.[58] The longer we overlook an invasive species, the tougher, costlier and more time-consuming the battle for control becomes. They are not waived off at our backyard fences, or at state, national or international boundaries. Awareness is one of the major campaign goals to stop the spread of this global menace and any states/countries have an exotic/invasive species council.

ENVIRONMENTAL MONITORS

In the natural chain, certain species are significant as indicators of environmental quality. Threatened species pose as our 'miner's canary',[59] i.e. their endangered state indicates at the growing imbalance in our life-support system. Measuring the health of an entire ecosystem is a huge task. Often, scientists look at one group of animals or plants as a metric of how well an ecosystem is working. For example, the swift decline in peregrine falcons and bald eagles in the mid-twentieth century was a vivid warning of the hazards of DDT.[60] The accumulation of this pesticide in the body of falcons and eagles hampered their fertility and egg-hatching process. Thus, many non-endangered species are used to monitor environmental quality.

Lichens, to a large extent, and certain plants like the eastern white pine can alert us to the accumulation of excess ozone.[61] Macrophytes like cattails, hydrilla, duckweed and water hyacinth are important monitors to access the accumulation of effluents, herbicides and industrial chemicals.

In another case, vultures are on the threshold of extinction in India because a banned drug is still being used unlawfully to treat suffering cattle. The endangered birds eat the carcass of the drugged animals and suffer kidney failure and visceral gout. The banned drug, diclofenac, remains a major concern, despite a long battle to wipe it out. Vultures maintain the cleanliness of our ecosystem by feeding on carcasses and hence prevent the spread of diseases to other animals and even humans. Nine species of vultures are found in India. Among them, 'four are critically endangered, one endangered, three near threatened and one least concern as per the IUCN status'.[62]

Freshwater mussels are very important environmental indicators. The eastern United States is home to the richest diversity of freshwater mussels in the world. These animals serve as 'filter feeders' as they draw in water and drain out food particles. This unique method of feeding helps to keep our waters clean. Their feeding habit, which proves beneficial for the ecosystem, attests harmful reports caused by none other than anthropocentric activities. As mussels filter particles from the water, they are habitually the first animals to be affected by water pollution. They tend to amass all kinds of toxins (chemicals in agricultural and industrial overspill) that are present in their habitat. Too much pollution can lead to the elimination of the mussels.[63] Today, most of the native freshwater mussel species are deemed endangered or of special concern. As such, without environmental monitors, we may not have learned of such perilous contaminants until much more damage had taken place.

Thus, Nature is an invisible yet invincible chain. Although nearly 1.20 million organisms of this chain have been identified by scientists, it is believed that an overwhelming majority has yet not been catalogued, and that almost 8.70 million species may inhabit an ecosystem.[64] There is an intricate and balanced correlation between each species, including us, the Homo Sapiens. This network is called the biosphere, which is, in turn, composed of numerous ecosystems, which include flora (plants) and fauna (animals) and their somatic environments. The vanishing of a single species can spark a chain reaction distressing many others. This is especially true for 'keystone' species,[65] which has an excessively large effect on its natural environment, and whose extinction can alter the ecological processes or fundamentally modify the species structure of the wildlife community.

POST-EXTINCTION ECOSYSTEM

Imagine a world without butterflies or the euphonious of the cuckoo, the dancing rivers, or the swarming fireflies. No forests to explore or mountains to scale. To biological conservationists, there is little hope in discussing an extinction, for mass extinctions literally wipe out everything. However, when looked at from an evolutionary perspective, extinction is a double-edged sword. As we understand it, extinction ends lineages and thus erases a variety of unique genetic codes and adaptations. But over vast geological timespans, it can redesign the evolutionary setting in multiple creative ways. This takes place through variance in survivorship of lineages and the evolutionary prospects occasioned by the extermination of dominant species. This is coupled by the post-extinction cataloging of survivors. To simplify: Extinction Level Events (ELE) engage in an interplay of the destructive and generative aspects of extinction. However, the different time scales over which they appear to operate remain a crucial but poorly understood cog of the evolutionary process.

Numerous series of evidence propose that post-extinction ecosystems function differently than those that exist before and long after the ELE. These entail important changes in geochemical and geological records which, in turn, indicate the upheaval in the global scale of various components of earth's ecology, such as the carbon cycle, corresponding with the extinction. As such, mass extinctions are not endings, but rather a stride into extended alternative ecosystems. Additionally, survivors of extinctions are an uncommon lot, demonstrating layers of morphological, ecological and physiological characteristics influenced by the selectivity of the extinction.

While researching, I came across an interesting term that signifies an important aspect of the post-extinction evolutionary trend. This is called the 'Lilliput Effect'.[66] Ever since I stopped growing beyond 5′3″, I was habitually teased by my brother with the epithet 'lilliput'. In the beginning we always ended up wrestling. As I grew up, the physical battles turned into verbal arguments supporting my dictum—lethal lilliput. Today, I find an even stronger argument to put across, one that has scientific ground. The 'Lilliput effect' is one of the permeating trends seen in the evolutionary history of many stocks resulting in size changes. Jonathan Swift's famous book *Gulliver's Travels*[67] mentions Lilliput as a fictional island inhabited by people 'somewhat under six inches high'—the Lilliputians. Modern-day paleontologists have embraced the term, 'the Lilliput effect' to mean something faintly altered. Organisms that survive mass extinctions tend to be much smaller than those that came before. Imagine the large size of marine fishes swarming the oceans today, as large as a crane or a bus. That's how it used to be almost 360 million years ago.[68] So, what might have occurred in the intermittent period? The answer lies in the massive extinction that happened about 359 million years ago, at the end of the Devonian Period. The scientists popularly named it the Lilliput effect, referring to a temporary size reduction following a mass extinction, but sometimes the temporary trend becomes permanent in specialized groups, such as plankton, birds or island fauna. The mechanism essential to the effect is controversial, with scientists putting forth multiple operational models based or weather, temperature, etc. Nevertheless, a recent study points to an important advantage of smaller size: smaller beings grow and reproduce quicker, and their adaptation to their environments is relatively fast because of their comparatively short generation times. The notion of 'bigger is better' developed by nineteenth-century paleontologist Edward Drinker Cope is definitely advantageous in the ability to fight off a predator or take advantage of a wide variety of resources. But large size and heftiness are advantageous as long as the going is easy. After a catastrophe, its contrasting 'Lilliput effect' prevails.

LET'S TURN THIS AROUND

Humans are increasingly usurping land that once belonged to the wild, which is directly proportional to the rapid dwindling of and final extermination of our bio-heritage. However, as ecologists stress, this need not alone cause their extinction. What is more alarming is the large number of humans wantonly disregarding the welfare of the wild (flora and fauna alike) and refuse cooperation with conservationists or even recognizing the idea of conservation itself. It's not the Darwinian sense of race, but selfishness, greed, shortsightedness, and lack of planning that have caused most contemporary species extinctions. In fact, a number of our game birds or animals would today be extinct were it not for cosseting through refuges and protective laws. Let us then draw together for a 'better' kind of thinking, living, and breathing keeping in mind that most of nature's contributions to the human community are not completely replaceable, and some are irreplaceable. Let us learn to co-exist, for the planet belongs to each one of 'us', humans, animals and plants alike. In the march of our 'progress', let's not create martyrs out of other species. For, the story of the sixth mass extinction will be a story of humans as much as it will be of birds, animals and trees.

RAISON D'ÊTRE

This volume is titled *Biological Apocalypse* since species extinction is a major factor of apocalyptic consequences. Tracing its etymology to Latin 'apocalypsis' (revelation) or Greek 'apokálypsis' (revelation, disclosing, uncovering), the term apocalypse has almost become synonymic with the age we live in—the Anthropocene or the Sixth Mass Extinction.

A note on the title of the Introduction seems imperative as well. The myopic and self-serving human world has already shoved the blue planet into the Sixth Mass Extinction phase. The Introduction takes into account different aspects of extinction ranging from its historical accounts, single species extinction, mass extinction and the debilitating invisible natural chain that binds the human and non-human world. Hence, considering our biosphere and the looming apocalypse, the portmanteau term 'biocalypse' has been conceived.

Biological Apocalypse consists of seventeen chapters which aim at inciting alarm in readers while inspiring them to take immediate action. Each chapter deals with a select species from the IUCN Red list which have been clustered into four sections: Critically Endangered, Endangered, Vulnerable

TABLE I.1: List of Species

Jerdon's Courser	Critically Endangered
Black Softshell Turtle	Critically Endangered
White-Bellied Heron	Critically Endangered
Chinese Pangolin	Critically Endangered
Oriental White-backed Vulture	Critically Endangered
Great Indian Bustard	Critically Endangered
Javan Rhino	Critically Endangered
Orangutan	Critically Endangered
Amur Leopard	Critically Endangered
Elephant	Endangered
Red Panda	Endangered
Water Buffalo	Endangered
White-Winged Wood Duck	Endangered
Yellow-throated Bulbul	Vulnerable
Indian Rhino	Vulnerable
Polar Bear	Vulnerable
Kaiser-I-Hind	Near Threatened

and Near Threatened. Further, replicating the natural chain and its interconnectedness, this book is interdisciplinary in nature with chapters cutting across species and their study: ornithology, mammalogy, lepidopterology, herpetology.

The chapters have methodical research-based content and points of view, with extensive textual and field research carried out over the years, substantiated with necessary references. The chapters extensively dwell on the anthropocentric reasons for the shrinking numbers of the species and the expanding Red List, apart from focusing on taxonomic specifics, a detailed report on the cause and effect of its decline, and suggestions to convalesce the loss.

The authors belong to different domains, ranging from scholars to scientists and activists. Hence, the quintessence of their outlook and the language exhibit an assortment of approaches to the species under study. However, the prime objective of each, tending to the purpose and design of the book, remains the same, i.e. to look at the micro-level catastrophes being induced by humans in the form of single-species extinction, and thereby to paint the larger picture of mass extinctions.

Entombed in our everyday is a narrative of multitude of silent exterminations. *Biological Apocalypse* throws light on not just the morbidity of the age that we are in, the Sixth Mass Extinction, but also attempts at instilling an appreciation in the readers on the marvels of the natural world.

NOTES

1. K. Jones et al., 'Global Trends in Emerging Infectious Diseases', *Nature*, vol. 451, 2008, pp. 990–3.
2. T. Crowther et al., 'Mapping Tree Density at a Global Scale', *Nature*, vol. 525, 2015, pp. 201–5.
3. Michael J. Zimmerman and Ben Bradley, 'Intrinsic vs. Extrinsic Value', *The Stanford Encyclopedia of Philosophy*, ed. Edward N. Zalta, 2019, see https://plato.stanford.edu/archives/spr2019/entries/value-intrinsic-extrinsic/, accessed 20 December 2022.

4. Makayla, 'Is it True That 99.9% of All Species Are Extinct?', 2019, see https://www.encyclopedia.com/daily/is-it-true-that-99-9-of-all-species-are-extinct/, accessed 20 December 2022.
5. George Cuvier, *Essay on the Theory of the Earth*, 1815; repr., Oxford: Oxford University Press, 2019.
6. D. Darwin Zenger, 'Lyell and Gradualism', *Nature*, vol. 322, 1986, p. 10.
7. Ashraf Elewa, 'The History, Origins, and Causes of Mass Extinctions', *Journal of Cosmology*, 2009, pp. 201–20.
8. Hannah Ritchie and Max Roser, 'Biodiversity', *OurWorldInData.org* (website), 2021, see https://ourworldindata.org/biodiversity, accessed 20 December 2022.
9. R. Silverberg, *The Auk, the Dodo and the Oryx*, New York: Thomas Y. Crowell, 1967.
10. B. Livezey 'An Ecomorphological Review of the Dodo (Raphus cucullatus) and Solitaire (Pezophaps soliaria), Flightless Columbiformes of the Mascarene Islands', *Journal of Zoology*, vol. 230, 1993, pp. 247–92.
11. Julian Hulme, 'The Dodo: From Extinction to the Fossil Record', *Geology Today*, vol. 28, no. 4, 2012, p. 148.
12. Ibid.
13. E.H. Bucher, 'The Causes of Extinction of the Passenger Pigeon', in *Current Ornithology*, vol. 9, ed. D.M. Power, Boston: Springer, 1992; D. McKinley, 'A History of the Passenger Pigeon in Missouri', *The Auk*, vol. 77, 1960, pp. 399–420.
14. G. Durrell, *Golden bats and Pink Pigeons: A Journey to the Flora and Fauna of a Unique Island*, New York: Simon and Schuster, 1977.
15. A. Cheke, 'The Legacy of the Dodo: Conservation in Mauritius', *Oryx*, vol. 21, no. 1, 1987, pp. 29–36.
16. Vaughan and P.O. Wiehe, 'Studies on the Vegetation of Mauritius. III. The Structure and Development of the Upland Climax Forest', *Journal of Ecology*, vol. 29, 1941, pp. 127–60.
17. S.A. Temple, 'Plant-Animal Mutualism: Coevolution with Dodo Leads to Near Extinction of Plant', *Science*, vol. 197, no. 4306, 1977, pp. 885–6.
18. B.K.P. Horn, 'Dodo Apocrypha', *Science News*, vol. 113, 1978, p. 19; A.W. Owadally, 'The Dodo and the Tambalacoque Tree', *Science*, vol. 203, 1979, pp. 1363–4; A. Cheke et al., 'Did the Dodo Do It?', *Animal Kingdom*, vol. 87, 1984, pp. 4–6; R.E. Vaughan, 'Did the Dodo Do It? Response to Temple', *Animal Kingdom*, 1984; Anthony Cheke, 'Extinct Birds of the Mascarenes and Seychelles: A Review of the Causes of Extinction in the Light of an Important New Publication on Extinct Birds', *Phelsuma*, vol. 21, 2013, pp. 4–19.

19. 'Why Save Endangered Species?', *U.S. Fish and Wildlife Service* (website), 4 April 2021, see https://www.fws.gov/sites/default/files/documents/Why-Save-Endangered-Species-Brochure_0.pdf, accessed 20 December 2022.
20. 'Houston Toad (Anaxyrus houstonensis)', *Texas Parks and Wildlife* (website), 4 April 2021, see https://tpwd.texas.gov/huntwild/wild/species/htoad/, accessed 20 December 2022.
21. Ross Piper, *Extinct Animals: An Encyclopedia of Species That Have Disappeared during Human History*, US: Greenwood Press, 2009, pp. 158–60.
22. Ibid.
23. M. Greshko and National Geographic Staff, 'What are Mass Extinctions, and What Causes Them?', *National Geographic Science*, 2019, see https://www.nationalgeographic.com/science/article/mass-extinction, accessed 20 December 2022.
24. Elewa, 'The History, Origins, and Causes of Mass Extinctions'.
25. M. Greshko, 'What Actually Killed the Dinosaurs? Volcanic Clues Heat up Debate', *National Geographic Science*, 2019, see https://www.nationalgeographic.com/science/article/what-actually-killed-dinosaurs-volcanoes-heat-up-debate, accessed 20 December 2022.
26. D. Carrington, 'Earth's Sixth Mass Extinction Event under Way, Scientists Warn', *The Guardian*, 2017, see https://www.theguardian.com/environment/2017/jul/10/earths-sixth-mass-extinction-event-already-underway-scientists-warn, accessed 20 December 2022; Niles Eldredge, 'The Sixth Extinction', *ActionBioscience* (website), 2001, see http://endangeredink.com/programs/population_and_sustainability/extinction/pdfs/Eldridge-6th-extinction.pdf, accessed 20 December 2022.
27. H. Trishler, 'The Anthropocene', N.T.M., vol. 24, 2016, pp. 309–35, see https://link.springer.com/article/10.1007/s00048-016-0146-3#citeas, accessed 20 December 2022.
28. Jeffrey Kevin McKee, *Spring Nature: The Conflict between Human Population Growth and Earth's Biodiversity*, New Jersey: Rutgers University Press, 2003.
29. Rima Kumari et al., 'Deforestation in India: Consequences and Sustainable Solutions', in *Forest Degradation Around the World*, ed Mohd Nazip Suratman et al., 2019, see https://www.intechopen.com/chapters/66710, accessed 20 December 2022.
30. Sandra Díaz, Josef Settele and Eduardo Brondízio, *Summary for Policymakers of the Global Assessment Report on Biodiversity and Ecosystem Services of the Intergovernmental Science-Policy Platform on Biodiversity and Ecosystem Services*, 2019, see https://ipbes.net/sites/default/

files/downloads/spm_unedited_advance_for_posting_htn.pdf, accessed 20 December 2022.

31. Mckee, *Spring Nature*, p. 3.
32. Yuval Noah Harari, 'Industrial Farming is One of the Worst Crimes in History', *The Guardian*, 2015, see https://www.theguardian.com/books/2015/sep/25/industrial-farming-one-worst-crimes-history-ethical-question, accessed 20 December 2022.
33. Navjot Sodhi, Barry Brook and Corey Bradshaw, 'Causes and Consequences of Species Extinctions', in *The Princeton Guide to Ecology*, Berlin: De Gruyter, 2009.
34. Harari, 'Industrial Farming is One of the Worst Crimes in History'.
35. Rachel Nuwer, 'What Would Happen if All the Worlds' Trees Disappeared', *BBC*, 2019, see https://www.bbc.com/future/article/20190911-what-would-happen-if-all-the-worlds-trees-disappeared, accessed 20 December 2022.
36. Ibid.
37. Jeff Berardelli, 'With More Species at Risk of Extinction, Study Warns of "Biological Annihilation"', *CBS News*, 2020, see https://www.cbsnews.com/news/species-extinction-risk-biological-annihilation-study/, accessed 20 December 2022.
38. Nanditha, *Hinduism and Nature*.
39. Ibid.
40. Ibid.
41. Rex Weyler, 'Gaia: Everything on Earth is Connected', *Greenspace.org* (website), 2019, see https://www.greenpeace.org/international/story/24978/gaia-ecology-earth-is-connected-rex-weyler/, accessed 20 December 2022.
42. Book of Genesis, see http://www.vatican.va/archive/bible/genesis/documents/bible_genesis_en.html.
43. Rudolf Simek, *Dictionary of Northern Mythology*, UK: D.S. Brewer, 1993.
44. Timothy Bourns, 'The Language of Birds in Old Norse Tradition', *Journal of English and German Philology*, vol. 12, no. 2, pp. 209–28.
45. Birds of the Bible: Solomon's Wisdom about Birds', *Lee's Birdwatching Adventures* (blog), 2013, see https://leesbird.com/2013/11/12/birds-of-the-bible-solomons-wisdom-about-birds-i/, accessed 20 December 2022.
46. V. Rajeev, 'Man and Nature in Indian Thought', *International Journal for Environmental Rehabilitation and Conservation*, vol. 4, no. 1, 2013, pp. 50–5.
47. Wordsworth, William and S.T. Coleridge, *Lyrical Ballads*, London: Penguin, 1786; repr., 2006.
48. Peter Wohlleben, *The Secret Network of Nature*, Kindle edn., Random House, 2017.

49. 'Wolves in Yellowstone: A Short History', *Yellowstone Insider*, see https://yellowstoneinsider.com/2009/05/03/wolves-in-yellowstone-a-short-history/, accessed 8 December 2022; Caeleigh Macneil, 'How Wolves saved the Foxes, Mice, and Rivers of Yellowstone National Park', *Earthjustice* (website), see https://earthjustice.org/blog/2015-july/how-wolves-saved-the-foxes-mice-and-rivers-of-yellowstone-national-park, accessed 20 December 2022; Katherine Lackey, 'Yellowstones Wolves are Back But They Haven't Restored the Parks Ecosystem: here's Why', *USA Today*, 7 September 2018, see https://www.usatoday.com/story/tech/science/2018/09/07/wolves-reintroduction-yellowstone-ecosystem/973658002/, accessed 8 December 2022.
50. M.E. Dorcas et al., 'Severe Mammal Declines Coincide with Proliferation of Invasive Burmese Pythons in Everglades National Park', *Proceedings of the National Academy of Sciences*, vol. 109, no. 7, 2012, pp. 2418–22.
51. Bharath Sundaram et al., 'Ecology and Impacts of the Invasive Species, Lantana camara, in a Social-Ecological System in South India: Perspectives from Local Knowledge', *Human Ecology*, vol. 40, 2012..
52. Rajat Rastogi, 'Lantana Invasion Threatens 40 Percent of India's Tiger Habitat, Reports Study', *Mongabay* (website), 2020, see https://india.mongabay.com/2020/08/lantana-invasion-threatens-40-percent-of-indias-tiger-habitat-reports-study/, accessed 20 December 2022.
53. Erin Spencer, 'Invasive Species Initiative: European Red Foxes', see http://www.invasivespeciesinitiative.com/european-red-foxes#:~:text=Introduction%3A%20Known%20for%20their%20luscious,be%20found%20throughout%20the%20country, accessed 4 April 2021.
54. University Of California-Davis, 'Invading Ants Disrupt Ecosystem', *ScienceDaily*, 2001, see www.sciencedaily.com/releases/ 2001/10/011011065354.htm, accessed 11 May 2020.
55. S. Lowe, M. Browne and S. Boudlejas, '100 of the World's Worst Invasive Alien Species', *Aliens*, vol. 12, 2000, pp. 1–12.
56. J. Drescher, N. Blüthgen and H. Feldhaar, 'Population Structure and Intraspecific Aggression in the Invasive Ant Species Anoplolepis gracilis in Malaysian Borneo', *Molecular Ecology*, vol. 16, 2007, pp. 1453–65.
57. Simon Devin and Jean-Nicolas Beisel, 'Biological and Ecological Characteristics of Invasive Species: A Gammarid Study', *Biological Invasions*, vol. 9, 2007, pp. 13–24.

58. Invasive Species', *National Wildlife Federation* (website), n.d., see https://www.nwf.org/Educational-Resources/Wildlife-Guide/Threats-to-Wildlife/Invasive-Species, accessed 5 March 2020.
59. Frank J. Mazzotti, 'The Value of Endangered Species: The Importance of Conserving Biological Diversity', *IFAS Extension*, 2017, see https://edis.ifas.ufl.edu/pdf/UW/UW06400.pdf, accessed 20 December 2022.
60. 'Peregrine Falcon (Falco peregrinus)', *U.S. Fish & Wildlife Service* (website), 2006, see https://ecos.fws.gov/ecp/species/8483, accessed 20 December 2022.
61. M.E. Conti, 'Lichens as Bioindicators of Air Pollution', *WIT Transactions on State of the Art in Science and Engineering*, vol. 30, 2008; Peter Reich et al., 'Effects of Ozone and Acid Rain on White Pine (Pinus strobus) Seedlings Grown in Five Soils. III. Nutrient Relations', *Canadian Journal of Botany*, vol. 66, 2011.
62. Hamim Mushtaq Wani et al., 'Status of Vultures in India: A Review', *International Journal of Advanced Scientific Research and Management*, vol .3, no. 12, 2018, p. 181.
63. 'Why Save Endangered Species?'.
64. Biodiversity', *National Geographic*, n.d., see https://education.nationalgeographic.org/resource/biodiversity, accessed 1 January 2022.
65. R.T. Paine, 'Food Web Complexity and Species Diversity', *American Naturalist*, vol. 100, 1966, pp. 65–75.
66. Peter Harries and Paul Knorr, 'What does "Lilliput Effect" Mean?', *Palaeogeography, Palaeoclimatology, Palaeoecology*, vol. 284, 2009, pp. 4–10.
67. Jonathan Swift, *Gulliver's Travels*, 1726; repr., New Delhi: Avenel Books, 1985.
68. Sid Perkins, 'Ancient Mass Extinction May Have Shrunk Earth's Creatures', *Science*, 2015, see https://www.science.org/content/article/ancient-mass-extinction-may-have-shrunk-earth-s-creatures, accessed 20 December 2022.

PART I

CRITICALLY ENDANGERED

1

Jerdon's Courser

An Enigmatic Indian Bird

Carol Inskipp

> Living wild species are like a library of books still unread. Our heedless destruction of them is akin to burning that library without ever having read its books.
>
> —John Dingell

Jerdon's courser (*Rhinoptilus bitorquatus*) could be India's most enigmatic bird. Known only from a small area of Andhra Pradesh, it is an extremely rare bird with very sporadic sightings. It is one of India's endemic bird species and has been assessed as Critically Endangered (CE), the highest category at risk of extinction in the IUCN list, with a decreasing population of 50–249 birds.[1]

Discovery

Jerdon's Courser was discovered in 1848 by a British physician and pioneering ornithologist, Thomas Caverhill Jerdon (1811–72), the civil surgeon to Nellore. The bird was

*The author is very grateful to P. Jeganathan for his most useful comments on the draft of this chapter.

first described from one of Jerdon's specimens and named after him by Edward Blyth,[2] who was curator of zoology at the Asiatic Society of India Museum. Jerdon discovered many bird species in India and several others bear his name: Jerdon's Bushchat, Jerdon's Baza, Jerdon's Leafbird and Jerdon's Babbler, to name a few.

In Telugu, the regional language of Andhra Pradesh, Jerdon's Courser is known as *Kalivi kodi*, because the bush under which the birds are found are called *Kalvi* chettu.[3]

APPEARANCE

Jerdon's Courser is an attractive bird in appearance, a little larger than the much more common and widespread Indian Courser (*Cursorius coromandelicus*), but unlike the latter, it has huge eyes, indicating its nocturnal habits. The adult has a broad buffish supercilium, narrow crown stripe, orange throat patch, and brown-and-white banding across the breast. Its legs are long and yellow. However, as regards the juveniles, it is still not known as to how they look.[4]

PLATE 1.1: Jerdon's Courser
Image courtesy: Bikram Grewal

Interestingly, Jerdon's Courser is much more closely related to the African courser than to the two coursers that are found in India: the Cream-coloured Courser (*C. cursor*) and the Indian Courser.

HISTORY OF RECORDS

In his detailed two-volume book, *The Birds of India*, published in 1862–72,[5] Jerdon described Jerdon's Courser as the Plover: '. . . This remarkable Plover has hitherto, I believe, only been procured by myself, from the hilly country above the Eastern Ghats, of Nellore, and in Cuddapah. It frequents rocky and undulating ground with thin forest jungle, and is found in small parties, not very noisy, but occasionally uttering a plaintive cry.' In May 1867 William Thomas Blanford saw the species 'close to Sironcha on the Godavari in eastern Maharashtra' and in March 1871 near Bhadrachalam in Andhra Pradesh.[6] These sightings by Blanford mark the known northern limit of the species' range and there are no further records from northern Andhra Pradesh.[7] In 1900, Howard Campbell recorded Jerdon's Courser near Anantapur in southern Andhra Pradesh.[8] This was the last known authentic sighting of the species until its sensational rediscovery nearly eighty-six years later. An enthusiastic Indian ornithologist, Bharat Bhushan, found it near Reddipalli, Cuddapah district, with the help of local tribal hunters.[9] As soon as the bird was re-discovered, the Andhra Pradesh government declared that entire area as Sri Lankamaleswara Wildlife Sanctuary.

BEHAVIOUR AND BREEDING

Very little is known about the behaviour of Jerdon's Courser. Although it can fly quite well, it is cursorial and prefers to walk and is active at night. Evidence for its diet is indicated by two samples of droppings that were found to consist of body parts of *Macrotermes* termites and ants.[10] Nothing is known about the bird's breeding ecology; no nests or young have been found. However, in 2008, an egg labelled as belonging to this species was found in an Aberdeen University collection and confirmed as coming from Jerdon's Courser using DNA. It was collected by Ernest Gilbert Smeaton, a veterinary surgeon, probably in 1917, although the locality is unknown. The egg is a short oval in shape, and pale yellowish with irregular brown blotches.[11]

CURRENT DISTRIBUTION

Since Jerdon's Courser was rediscovered, a number of surveys have been carried out to determine its current distribution. However, it has only been found in a few restricted areas in Cuddapah district in southern Andhra Pradesh.[12] There have been few actual sightings since the bird's rediscovery; for example, Indian ornithologist P. Jeganathan who worked on the bird for several years, only saw it a handful of times.

SURVEY METHODS

The courser is very difficult to locate; it naturally occurs in low densities; it is nocturnal and its calls are not frequently heard. One reliable way to find it is to set up about twenty or thirty camera traps in a potentially suitable habitat for about a month. If the birds are there, and if one is lucky, they might trigger a camera.[13] As the birds are so elusive, another method for detecting their presence was innovated in 2001. Several 5 m. long tracking strips of fine-grained soil about 2 cm. thick were laid in known courser habitat. Automatic cameras placed at the end of some of these tracking strips helped to identify the bird that had left footprints on the soft strips. It was soon possible to distinguish Jerdon's Courser footprints from those of other birds.[14] Surveys using this method found the bird in three new places.[15] The call of the bird has also been used to locate new sites. In 2002 a Jerdon's Courser was observed calling which was recorded on tape. A tape transect method was devised and using this tape was played at 403 points; the courser responded at three of these. All these sites were within 14 km. of the discovery site.[16]

HABITAT

The specific habitat of Jerdon's Courser was identified from the above described survey methods. Unlike the other two courser species in India, which inhabit open areas, Jerdon's Courser was found to prefer areas with woody plant species such as *Hardwickia binate*, which is characteristic of mature scrub that has not been cleared or suffered a major disturbance for a long time.[17] Satellite imagery has been used to map the extent of suitable habitat which has scrub densities preferred by Jerdon's Courser. Unfortunately, only after 2000, based on field research, was it found that its potentially suitable habitat lies at the foothills of the Lankamalai Hill ranges (the Eastern Ghats) and most of these places were outside the protected area.[18]

THREATS

Loss and degradation of scrub forest has been found to be the major threat to Jerdon's Courser.[19] In 2005, the construction of the Telugu Ganga Canal, a large irrigation canal, destroyed some of these newly discovered sites and probably led to several ecological changes to the scrub jungle habitat. When construction started near the only known location where Jerdon's Coursers are known to live, conservationists, forest department officials, and many organizations such as the Bombay Natural History Society (BNHS) and the World Wildlife Fund (WWF) India opposed it. Several conservation organizations and individuals wrote to the Andhra Pradesh government. Meetings were held and a Public Interest Litigation was filed. The Supreme Court of India asked BNHS to suggest an alternative route for the canal, and in 2008 the canal was rerouted from its original alignment to safeguard the remaining potentially suitable habitats of Jerdon's Courser.[20]

However, pressures on land use have continued, with swathes of scrub jungle at the fringes of the Sri Lankamaleswara Wildlife Sanctuary being cleared for agricultural expansion and monoculture plantations.[7] Livestock grazing, disturbance and quarrying are additional threats. Ground surveys and remote sensing have both shown there has been substantial clearance of scrub jungle near the outskirts of Sri Lankamaleswara Wildlife Sanctuary including in areas where there were recent records of the courser.[21]

Very sadly, there have been no confirmed records of the courser since 2008, only twenty-two years after it was rediscovered, according to P. Jeganathan who heard it's call in April 2008.

RECOMMENDATIONS FOR CONSERVATION

Fortunately, hope hasn't faded away yet as the BNHS and partners have developed a Species Recovery Plan with the Andhra Pradesh Forest Department. Recommendations for

PLATE 1.2: Camera-trap image of the Critically Endangered Jerdon's Courser In Sri Lankamaleswara Wildlife Sanctuary, Kadapa, Andhra Pradesh, 2001. *Image courtesy*: © P. Jeganathan

Jerdon's Courser include protection of all sites where it has been recorded since 2000 and, importantly, protection of any site with suitable habitat for the courser, even if the bird has not been found there yet. Protection, restoration and improvement of scrub jungle within the Sri Lankamaleswara Wildlife Sanctuary are recommended. Within the sanctuary, non-native plantations such as eucalyptus and biodiesel plantations should be removed and the areas restored to native scrub jungle. Illegal bird hunting and trapping should be stopped in the critical habitat, both inside and outside the sanctuary, as although the courser is not targeted, it could still be killed accidentally. The extent of native scrub jungle suitable for the courser should be mapped and measured at and around recorded courser sites using ground surveys and remote sensing every five years. It is also very important that publicity materials such as posters, stickers and brochures explaining the importance of Jerdon's Courser conservation should be produced in English and in Telugu. Awareness programmes should be increased in villages around the sanctuary and at state and national level to raise the profile of the species and gain support for conservation of this iconic bird.[22] We simply cannot afford to add another species in the list of extinction.

Since 2009 use of more efficient, digital camera traps have replaced film cameras in Jerdon's Courser research. As well as detecting the courser, they can generate a huge amount of information on habitat use by other wildlife, livestock grazing pressures and on human impacts such as fuel wood collection.[23]

More survey efforts are needed to find/sight this extremely elusive bird again, both in the areas where it was last seen and in areas of potentially suitable habitat which still exist across southern Andhra Pradesh.[24]

From all over the world, we hear similar tales of woe all the time. It is initially disappointing, and eventually just numbing. On the face of it, there are ample of reasons why we shouldn't bother saving vanishing species. However, the simple fact that the natural world is an invisible chain, placing even humans in its delicate balance, calls for a timely response and action towards the imminent extinction of any species. Each species is an integral constituent of an ecosystem, and its value assessment in terms of the services it provides should be a customary part of ecosystem assessments. However, since it is difficult to do this, value assessment shall always remain incomplete. Nonetheless, nature is beautiful, and that aesthetic value is itself a reason enough to protect it. We look forward to a day when Jerdon's Courser, a flagship species, finds no mention in the list of critically endangered species.

NOTES

1. 'Species Factsheet: *Rhinoptilus bitorquatus*', *Bird Life International* (website), 2020, see http://www.birdlife.org, accessed 12 May 2020.
2. E. Blyth, 'Report of Curator, Zoological Department', *J. Asiatic Society of Bengal*, vol. XVII, 1848, p. 254.
3. A. Gupta, 'Jerdon's Courser: A Precious Rarity', *India's Endangered* (website), 1 July 2011, see http://indiasendangered.com/jerdon%E2%80%99s-courser-a-precious-rarity/, accessed 12 May 2020.
4. R. Grimmett, C. Inskipp. and T. Inskipp, *Birds of the Indian Subcontinent*, Helm Field Guides, London: A&C Black, 2011.
5. T.C. Jerdon, *The Birds of India, being a Natural History of All the Birds Known to Inhabit Continental India; with Descriptions of the Species, Genera, Families, Tribes, and Orders, and a Brief Notice of such Families as are not Found in India, Making it a Manual of Ornithology Specially Adapted for India*, vol. II, part II, Calcutta: T.C. Jerdon, 1864.
6. W.T. Blanford, *The Fauna of British India*, vol. IV, London: Taylor and Francis, 1868.
7. P. Jeganathan, 'Jerdon's Courser', *Nature in Focus: Nature Conservation Foundation* (website), 2017, see https://www.natureinfocus.in/nature-and-wildlife-conservation/jerdon-s-courser, accessed 12 May 2022.
8. E.C.S. Baker, *The Nidification of birds of the Indian Empire*, London: Taylor and Francis, 1929.
9. B. Bhushan, 'Rediscovery of the Jerdon's or Double-banded Courser

Cursorius bitorquatus', *Journal of the Bombay Natural History Society*, vol. 83, 1986, pp. 1–14.

10. P. Jeganathan et al., 'Conservation of the Critically Endangered Jerdon's Courser *Rhinoptilus bitorquatus* in India', Mumbai: Bombay Natural History Society, 2004, p. 32.
11. A. Knox and S. Piertney, 'The Discovery of the Egg of Jerdon's Courser *Rhinoptilus bitorquatus* (1848) in India', *Journal of the Bombay Natural History Society*, vol. 103, no. 2, 2013, pp. 149–52.
12. Jeganathan, 'Jerdon's Courser'; A.R. Rahmani, *Threatened Birds of India: Their Conservation Requirements*, Mumbai: Bombay Natural History Society, 2012.
13. Jeganathan, 'Jerdon's Courser'.
14. P. Jeganathan et al., 'Use of Tracking Strips and Automatic Cameras for Detecting Critically Endangered Jerdon's Courser *Rhinoptilus bitorquatus* in Scrub Jungle in Andhra Pradesh', *Oryx*, vol. 36, 2002, pp. 182–8.
15. Rahmani, *Threatened Birds of India*; P. Jeganathan, 'Ecology of the Jerdon's Courser *Rhinoptilus bitorquatus* in India', PhD thesis, University of Mumbai, 2006.
16. Rahmani, *Threatened Birds of India*; Jeganathan, 'Ecology of the Jerdon's Courser'.
17. P. Jeganathan et al., 'Modelling Habitat Selection and Distribution of the Critically Endangered Jerdon's Courser *Rhinoptilus bitorquatus* in Scrub Jungle: An Application of a New Tracking Method', *Journal of Applied Ecology*, vol. 41, 2002, pp. 224–37.
18. Jeganathan, 'Jerdon's Courser'.
19. Ibid.; D. Senapathi et al., 'Use of Remote Sensing to Measure Change in the Extent of Habitat for the Critically Endangered Jerdon's Courser *Rhinoptilus bitorquatus* in India, *Ibis*, vol. 149, no. 2, 2007, pp. 328–37.
20. Jeganathan, 'Jerdon's Courser'; Rahmani, *Threatened Birds of India*.
21. Rahmani, *Threatened Birds of India*.
22. Ibid.
23. Ibid.
24. Jeganathan, 'Jerdon's Courser'.

2

From Darkness to Light
Saving the Black Softshell Turtle from Extinction

Shailendra Singh and *Rishika Dubla*

> We are in the midst of the 6th largest extinction event in the history of the plant and the first caused by human action.
>
> —GRACIELA CHICHILNISKY

THE HIDDEN CRISIS

Mediocrity is a misfortune suffered by the abundant. Something that dares to be common jades the roving human eye, which only seeks that which is rare, singular, hidden, and ensconced in decades of legend, for with it lies the glory of discovery. The natural world too is a victim of this ideology. The hunt for the extraordinary has been the zeitgeist of explorers since humans first began to study the natural world. Fauna and flora regularly encountered are rendered banal, their commonness taken for granted, and considered not worthy of our awe or even our attention, eventually playing background set pieces to centre stage species. One such set piece, outshined by its more charismatic cousins, is turtles. Even though India has one of the highest rates of chelonian biodiversity in the world, time and time again one hears the same story, where this abundance is taken for granted until it is too late. Such is the story of the Black Softshell Turtle (Nilssonia nigricans). Also known as the Bostami Softshell Turtle,[1] this species was once possibly widespread across its range of Bangladesh and north-east India.[2] But like the curse that has today claimed most turtle species, populations have dwindled to the point where it was officially declared as extinct from the wild (IUCN).

ORIGINS AND ADAPTATIONS

Turtles, also known as chelonians (an all-encompassing term that refers to marine and freshwater turtles, tortoises and terrapins), form an ancient group, dating back over 200 million years[3] to the Triassic Period. From this standpoint, turtles and tortoises walked the Earth when the continents of the world were still concentrated into the single supercontinent named 'Pangea',[4] where many of the popular dinosaurs we know today did not even exist.[5] They inhabit

PLATE 2.1: A Black Softshell Turtle in Nagshankar Temple Pond (Assam). *Image courtesy*: Shailendra Singh

marine, freshwater and terrestrial ecosystems, alike, often serving as keystone species.

The chelonian shell is one of the most recognized feature of the species; the only vertebrate in the entire history of life on the planet to have evolved such an adaptation.[6] However, it barely scratches the surface, where evolutionary traits such as delayed sexual maturity, extended reproductive lives and extreme longevity,[7] have allowed turtles and tortoises to survive for millions of years. Ironically, it is these very traits that now threaten their survival, jeopardized by another species that has only been around for a geological split second—humans.

DWINDLING NUMBERS

Today, 360 species of turtles are recognized worldwide, out of which nearly 61 per cent are threatened.[8] Of the 29 species of freshwater turtles found in India, 22 are listed as threatened according to IUCN Red List Assessments,[9] representing 75 per cent of Indian chelonian fauna. Globally, India ranks third in terms of countries with the highest number of threatened turtle species, with the top five endangered Indian species being represented in both the world's top twenty-five and top fifty lists of most threatened turtles.[10] N. nigricans is also part of this list, registered as 'Rare' since 1982. However, conservational apathy to safeguard the species expedited its decline to eventually being listed as 'Extinct in the Wild' in 2002.[11]

EMINENT DANGERS FROM APEX PREDATORS

Direct and indirect anthropogenic activities endanger turtles at every single life stage, offering no respite for a species to adapt to the onslaught of increasing environmental challenges, and the Black Softshell is no exception. Nesting beaches have been eradicated to the point of ecological unviability through development and hydroelectric projects, logging, clandestine sand collection and riverside agriculture in larger parts of its habitat. While the reproductive strategy of high fecundity can absorb predation pressures from natural predators such as jackals, dogs and birds, it is not able to handle the added pressure of gastronomic profligacy by humans. Removal of adults from an ecosystem occurs in numbers that are difficult for scientists, let alone the layman, to fathom: tens of thousands either die as accidental bycatch due to unsustainable fishing practices, or are hunted for local consumption, traded in wet markets, poached for national meat markets or illegally trafficked across porous borders into South-East Asian meat and traditional Chinese medicine markets. Exacerbated climate change through other human activities further puts them at risk, as most turtles, like other reptiles, rely on environmental sex determination. Rising temperatures skew sex ratios of the few surviving natural nests towards more females, further jeopardizing already declining populations.

LIGHT AT THE END OF THE TUNNEL

At its eleventh hour, it was the species' elevated status in both Hindu and Islamic mythology that came to its rescue against complete annihilation. A semi-captive population of roughly 450 individuals in the Bayazid Bostami Shrine in the Chittagong Hills is the last documented population of the species in Bangladesh;[12] it is protected hitherto by a local legend which claims that the turtles are descendants of evil spirits that were trapped in the shrine's pond by the saint Bayazid Bostami. Today, they are venerated as sacred animals. Further, in a groundbreaking study in 2007,[16] the species was discovered to still exist in the temple ponds of Assam in India, built by the Ahom Kings in the twelfth century AD for ritual bathing and drinking. These rulers also donated turtles as religious offerings to temple ponds to appease the gods, a practice still followed today by the people of Assam.[13]

Then, much to the delight of turtle biologists across the world, reports of wild occurrences began to trickle in,

PLATE 2.2: Nagshankar Temple Pond (Assam) houses sizeable populations of the Black Softshell Turtle. *Image courtesy*: © Shailendra Singh

from both Bangladesh and India.[14] Conservationists began to herald the temple populations as the ace in the hole that could help wild populations bounce back.

THE FLAWED MASTER PLAN

The primary conservation strategy adopted was to establish husbandry improvement led conservation breeding program in collaboration with temple authorities in the temples themselves, with an aim to release animals sourced from these captive populations back into the wild. Pilot projects run by the Turtle Survival Alliance India, Assam Forest Department and partners seems to be successful, releasing some hatchlings into the wild and shifting other animals into better facilities.[15]

On the surface, it was a brilliant plan that provided a promising view of cross-factional collaboration that made donors vie amongst each other to sponsor these projects. But this is where the hubris of scientists underestimated the

PLATE 2.3: Bostami Turtle Pond, Chittagong, Bangladesh.
Image courtesy: © Shailendra Singh

centuries-old power of deep-rooted religious sentiments and traditions. Again, in a cruel twist of fate, the sacred status of the Black Softshell that saved the species from extinction is now proving to be the very reason of its mounting existential crisis.

Today, the temple ponds are overcrowded, and animals are malnourished, riddled with disease and subject to infighting, often leading to cannibalistic behaviour. The ponds in Assam seldom have any basking platforms or appropriate nesting beaches, resulting in very poor nesting and hatching success rates. Even with the eggs that do hatch, predatory African catfish species that also reside in the ponds often prey upon hatchlings and juveniles. However, despite several attempts to convince temple authorities to release or shift these animals, they continue to be staunchly protective, refusing most such measures.

After nearly a decade of mediation between temple authorities, conservationists, stakeholders and Forest

PLATE 2.4: Kamakhya Temple Pond, which has a good breeding colony.
Image courtesy: © Shailendra Singh

Department officials, we have only gotten so far as convincing these temples to install basic husbandry upgrades and head-starting of new hatchlings in alternative community ponds. However, once big enough to survive predation from the catfish and other, bigger cannibalistic turtles, majority of hatchlings are immediately put right back into the temple ponds. However, more often than not, any minimal amount of progress slips back like footsteps in a sand dune due to the waning interest of stakeholders and the absence of sustained funding. This is exemplified by the work done by the TSA India Program who, for a short time, convinced priests from the Hayagriva Madhava and Nagshankar Temples to shift some animals into rearing enclosures; however, the temple authorities and local religious groups eventually changed their minds and, after the project team had left, promptly shifted the animals back into the main temple pond. But in last two years (2020–1) over 140 animals were translocated from Nagshankar to the nearest conservation facility—the

Nature Discovery Centre near Bishwanath—for short term rearing and release. The Sisyphean nature of these efforts is further nullified by the fact that temple authorities still continue to accept turtle donations, which, we believe, further contributes to clandestine turtle trade.

Another facet that stymies the conservation of the species is the administrative structure that governs these temples. Though the Forest Department is the custodian of all wildlife and forest-related issues, and these animals technically come under the ownership of the state administrations, temples have their own administrative committees that are slightly 'disjunct'. Thus, the Forest Department seldom finds a way to be part of this 'eco' system. Often, administrations are skeptical of sending strict advisories to these temples lest they incur their wrath, either as curse from the priests or through more Machiavellian exclusions from important inner political circles.

The folly lies in the belief that just because these animals are easily available, they are the Holy Grail to save the species. While the idea is not without its merits, most conservationists today have failed to consider that changing the mindset of the temple authorities to sanction release programs may take years, if not decades—time that the species does not have. In the short run, shifting primary focus from superficial release programs to more intensive efforts to safeguard the last remaining wild individuals and addressing fundamental administrative and policy issues outside the confining dicta of the temples may create longer-lasting conservation success for the species. It is time that there is a complete overhaul of the conservation strategy currently being implemented, where temple conservation serves as a mere cog in the whole system, rather than the entire system itself.

CHANGING CONSERVATION STRATEGIES

First and foremost, as attractive as direct conservation outputs via release programs may seem, it is time that donors realize the importance of conducting comprehensive surveys and more intensive studies on the population status and distribution of the species. This lack of formal surveys has plagued assessments from the beginning, where its population status, reproductive behaviour and distribution has been determined by the Bostami Shrine animals,[16] though as recent evidence suggests, individuals of N. nigricans may possibly exist in the wild. Better understanding of populations and their distributions may not only provide a better frame of reference to protect key areas, but might also allow scientists to evaluate key areas for reintroduction and species recovery.

Second, it is imperative to establish inclusion of government authorities that are constitutionally responsible for the welfare of threatened Indian fauna into the temple ecosystem, which, at the moment, have no formal involvement in the matter. We suggest that a central system that functions through the chief secretary of the state should allow for a better monitoring system for the welfare of the species, which is currently lacking a designated spokesperson

PLATE 2.5: A hatching cum rearing facility under construction at Nagshankar Temple (Sonitpur-Assam). *Image courtesy*: © Shailendra Singh

to even identify pressing issues. They can also issue general husbandry circulars such as 'White paper' to temples to ensure that the animals in captivity are being provided with the best quality of life, following national and international animal welfare and conservation guidelines.

Two of the major conservation challenges for the release program are lack of a formal release protocol and the risk of hybridization in the ponds. First, without quarantine measures, health assessments, size-class based survival studies and post-release monitoring protocols, there is a high risk of releasing diseased animals that could jeopardize healthy wild populations. Second, with other species of the same genus, such as the Indian Softshell Turtle (Nilssonia gangetica) and its sister species the Indian Peacock Softshell Turtle (Nilssonia hurum) inhabiting the same pond, there is suspicion of hybridization (cross-breeding), warranting robust genetic studies and bringing in new captive animals from other colonies. Before both these issues are dealt with, releasing animals, either bred or sourced from these captive ponds, may do more damage than good to wild populations.

This is not to say that the idea of temple conservation is not ideal; the concern is that the inspiration which has been distilled through a scientist's ideology is often antagonistic to a priest's, resulting in glacial attitudinal changes over the years. In reality, the heart of the matter lies in not trying to make temple priests understand our motivations for saving the Black Softshell, but to actually discover theirs. For example, it was only after a poacher attack in 2006 at the Bostami Shrine, where they poisoned the waters in the cover of night, that led the shrine authorities to improve security and employ 'khadems' (turtle guardians/caretakers). Water conditions and actively monitored nests significantly increased the hatching success of the Bostami turtles.[17] In another success story, by closely working with Pranab Malakar, a turtle pond caretaker at the Madhav Temple in Assam, the TSA India Program was able to make considerable improvements to husbandry and infrastructure at the temple. Basing our goals and subsequent actions within a context that is palatable to the priests may not only increase their acceptance of more conservation interventions on temple premises, but may even offer a chance in convincing them to freeze the intake of turtle donations. All such temple colonies may need to work together under the guidance of specialized conservation agencies and learn from each other's mistakes, while preparing local leaders such as Pranab could change the attitude and perception of other committees.

A MIND FRAME RESET

The world of conservation is rife with moral paradoxes. It is uncomfortable to think that the very adaptations that shot turtles towards the highest echelons of evolutionary success proved to be their downfall in the hands of human. But to absolve ourselves of this guilt, conservationists have, from the very moment that this field was birthed, taken a misguided stand of 'for the species, against the people'. And it continues to be the case with Black Softshell Turtles, where this attitude has struggled to show any success with temple collaborations and sidelined other viable options. As the pendulum of survival for the Black Softshell Turtle sways back and forth, earlier generations may have faced the moral quandary of determining where the importance of saving a threatened species ends and the value of human existence begins. But today, we must first ask ourselves whether these are even mutually exclusive. This could serve as an appropriate place to start with.

NOTES

1. G.S.M. Asmat, 'Nilssonia nigricans', *IUCN Red List of Threatened Species* (website), 2002.
2. S. Singh, S. Badola and M. Fernandes, *Identification Cards: Tortoises and Freshwater Turtles of India*, India: TRAFFIC, 2019.

3. C.H. Ernst and J.E. Lovich, *Turtles of the United States and Canada*, 2nd edn., Baltimore: Johns Hopkins University Press, 2009; S. Singh, *Impact Report: Disney Conservation Fund*, 2019, unpublished report.
4. M. Bagley, 'Triassic Period Facts: Climate, Animals and Plants', *Live Science* (website), 2014, see https://www.livescience.com/43295-triassic-period.html#:~:text=The%20Triassic%20Period%20was%20the,the%20oceans%20began%20to%20diversify, accessed 20 December 2022.
5. Turtle Conservation Fund, *A Global Action Plan for Conservation of Tortoises and Freshwater Turtles. Strategy and Funding Prospectus, 2002–2007*, Washington, DC: Conservation International and Chelonian Research Foundation, 2002.
6. E.J. Lovich et al., 'Where Have all the Turtles Gone and Why Does it Matter?', *BioScience*, vol. 68, no. 10, 2018.
7. Craig B. Stanford et al., *Turtles and Tortoises are in Trouble*, Current Biology, vol. 30, 2020, R721–R735.
8. Ibid.
9. Singh, Badola and Fernandes, *Identification Cards.*
10. B. Chatterjee, 'India has 10% of World's 50 Most Threatened Freshwater Turtle Species: Report', *Hindustan Times*, 2018, see https://www.hindustantimes.com/mumbai-news/india-has-10-of-world-s-50-most-threatened-freshwater-Turtle-species-report/story-srgbMsjH2lp9jxjTTXK1XL.html, accessed 20 December 2022.
11. Asmat, 'Nilssonia nigricans'.
12. M.F. Ahsan, M.N. Haque, and C.M. and Fugler, 'Observations on Aspideretes nigricans, a Semi-domesticated Endemic Turtle from Eastern Bangladesh', *Amphibia-Reptilia*, vol. 12, 1991, pp. 131–6; N. Annandale and M. Shastri, 'Relics of the Worship of Mud Turtles (Trionychidae) in India and Burma', *J. Proc. Asiatic Soc. Bengal* (n. s.), vol. 10, 1914, pp. 131–8; Subhashish Das et al., 'Conservation Status of Black Soft Shell Turtle (*Nilssonia nigricans*) in Bangladesh', *Journal of Wildlife Research*, vol. 6, no. 2, 2018, pp. 12–16; B. Groombridge, *The IUCN Amphibia-Reptilia Red Data Book*, part 1: Testudines, Crocodylia, Rhynchocephalia, Gland: IUCN, 1982.
13. Dewani, U., 'Forgotten Waterbearers of Guwahati', *India Water Portal* (website), 2016, see indiawaterportal.org/articles/forgotten-water-bearers-guwahati; D.G. Niyogi, 'Moved from Temple Ponds to the Wild, black Softshell Turtles Walk towards Revival', *Mongabay* (website), 2020, see https://india.mongabay.com/2020/05/moved-from-temple-ponds-to-the-wild-black-Softshell-Turtles-walk-towards-revival/, accessed 20 December 2022.

14. M.F. Ahmed and A. Das, 'Tortoises and Freshwater Turtles of Kaziranga National Park, Assam: Diversity, Distribution, Conservation Status', in *Freshwater Turtles and Tortoises of India*, ENVIS Bulletin, vol. 12, no. 1, 2009, pp. 57–70; N. Liebing et al., 'Molecular Phylogeny of the Softshell Turtle Genus Nilssonia Revisited, with First Records of N. formosa for China and Wild-living N. Nigricans for Bangladesh', *Vertebrate Zoology*, vol. 62, no. 2, 2012, pp. 261–72; Singh, Badola and Fernandes, *Identification Cards*.
15. Singh, *Impact Report*; Niyogi, 'Moved from Temple Ponds to the Wild, black Softshell Turtles Walk towards Revival'.
16. Ahsan, Haque, and Fugler, 'Observations on Aspideretes nigricans'; M.F. Ahsan and M.A. Saeed, 'The Bostami Turtle, Trionyx nigricans Anderson: Population Status, Distribution, Historical Background and Length-weight Relationship', *Bombay Natural History Society*, vol. 86, 1989, pp. 2–5; Annandale and Shastri, 'Relics of the Worship of Mud Turtles (Trionychidae) in India and Burma'.
17. Das et al., 'Conservation Status of Black Soft Shell Turtle (*Nilssonia nigricans*) in Bangladesh'.

3

White-Bellied Heron

A Rare, Critically Endangered Ave

Sarala Khaling

> Birds are indicators of the environment. If they are in trouble, we know we will soon be in trouble.
>
> —Roger Tory Peterson

There are very few birds in the world which are as rare and threatened as the White-bellied Heron (WBH). The species is known to have a large range, from Bhutan in the west through Assam and Arunachal Pradesh in India to Myanmar in the east and, more lately, probably beyond the Himalayas in Yunan province, China. It was found as far west as Bihar, Nepal and north Bengal. However, even though the area of their distribution looks large, they are found in only a few pockets within this range. In India, many of the areas that could house the bird have not been surveyed by scientists and researchers. These include remote areas, with difficult terrain, dense forests and swift-flowing rivers. Surveying these areas requires time and financial resources which are not easily available. This could be one of the biggest threats to the bird in India. Northeast India is changing rapidly and in order to attain economic growth hydropower dams, highways and smaller roads, bridges and other infrastructure are mushrooming everywhere. With the increase in population towns and villages are growing, which are often detrimental to natural areas if they are not planned well. Tea plantations, rubber and oil palm plantations are also increasing rampantly in private land at the cost of natural forests in north-east India. Many of these places are unexplored and we may therefore be missing out on the species before even establishing their presence at a particular site. The priority now is to survey all potential habitats where the birds may be present without which we may lose key habitats without even knowing if the bird was found there.

TAXONOMY

The WBH was described first by Allan Octavian Hume, who was in the Imperial Civil Service of the British Empire in India. Among other things, he was a well-known ornithologist and botanist. For his work on birds, he has often been referred to as 'the Father of Indian Ornithology'. WBH was described by Hume from a specimen collected from the then Sikkim Terai and/or the Bhutan Duars.

The great Indian ornithologist Salim Ali and the American ornithologist and wildlife conservationist Sidney Dillon Ripley called WBH the Great White-bellied Heron due to its large size. The same nomenclature (naming) was also given by two other ornithologists from the UK, Nigel Collar and Paul Andrew, who named it the Imperial Heron (*Ardea imperialis*). The accepted name now is the White-bellied Heron (*Ardea insignis*).

SYSTEMATICS

Order	Pelecaniformes
Family	Ardeidae
Genus	Ardea
Scientific Name	*Ardea insignis*

SPECIES DESCRIPTION

The WBH is a large bird and the second largest (the first being the Goliath Heron) living species of heron in the world. It has a height of about 127 cm. and a large blackish bill which is 15–18 cm. long. It is mostly dark greyish in colour with a contrasting white throat, belly and vent and white-streaked scapulars/shoulders, fore neck and upper breast. The legs and feet are grey. Adult males and females have two lace-like white plumes on their nape, while juveniles have smaller browner plumes.[1]

DISTRIBUTION

Historically the WBH was found in Nepal, Bihar and north Bengal, where it is now considered to be locally extinct. On a map, its distribution is found mainly on the southern side of the eastern Himalayas (eastern Nepal–north-west Myanmar).

PLATE 3.1: A White-bellied Heron
Image courtesy: Yumlam Benjamin Bida/ATREE

It is resident in Bhutan, India and Myanmar. There were two observations of the bird by local people in a rural area in Yunan province of China, which means its range may extend east of the Himalayas as well. Wherever it is distributed, the species is found in low numbers and the overall population is regarded as insufficient for future long-term survival.[2] In India, this species occurs in north-east India in two states—Assam and Arunachal Pradesh. In Assam it is found in the Manas area adjoining Bhutan in the north. It does not stay on the Indian side but we think it comes to feed on the river bank. The Namdapha Tiger Reserve in Arunachal Pradesh is the main place where the species is found.[3] Scientists and other observers estimate five–six individuals thus far but many places in Namdapha which were not surveyed were then studied and we found new places that were the birds' natural habitats. We can now say that there may be about 9–10 birds in Namdapha. In early 2019, for the first time the WBH was also recorded in Kamlang Tiger Reserve, in the Lohit district of Arunachal Pradesh.[4] In 2021 the White-bellied Heron was reported for the first time in the pine forests of Walong in Anjaw district of Arunachal Pradesh.[5]

HABITAT

The WBH lives, feeds, breeds and roosts in areas along rivers in forests, wetlands, grasslands, in small or large rivers usually with sand or gravel bars, and in areas covered in cobblestones and pebbles, often within or adjacent to subtropical broadleaved forest, and more recently in pine forests (*Pinus merkusii*). It is usually found from the lowlands up to at least 1,500 m.[6] In China it was found in a place by local people at an elevation of 2,100 m. In Myanmar, observers have seen the bird in rivers with beds of 'large shingle stones' at elevations of 110 m and 300 m.[7] In Bhutan they are also found around large rivers and lakes with Chir pine trees (*Pinus roxburghii*).

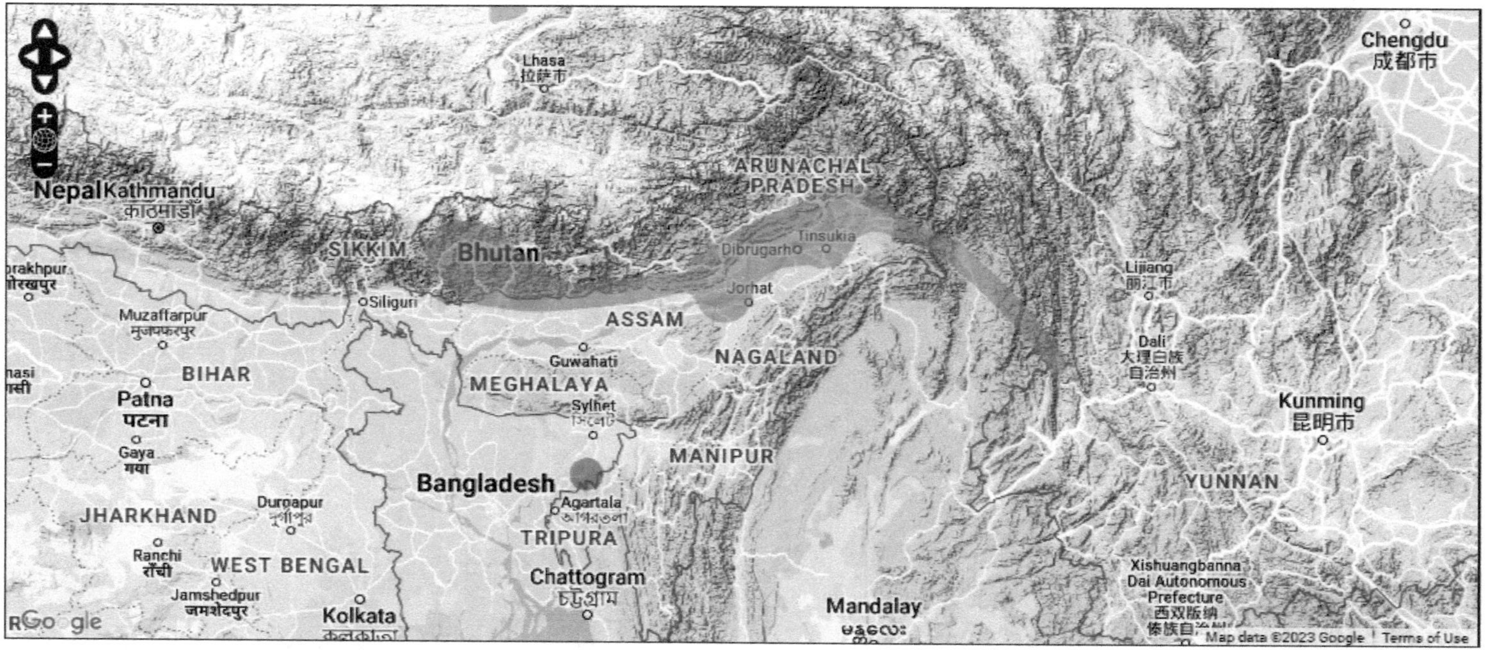

The international boundaries of India as shown in the map are neither correct nor authentic.

MAP 3.1: Distribution map of WBH
Image courtesy: Birdlife International 2023

FORAGING AND FEEDING BEHAVIOUR

The WBH feeds on fishes and is very good at catching them. It feeds alone or in pairs, depending on the season. Feeding in pairs possibly takes place during its courtship season (last week of February) and after the fledgling of chicks (July–August).[8] WBH mostly stands and waits for a long time along fast-flowing rivers to catch fishes. It also walks slowly along streams and grassy marshes looking for fishes.[9] The bird seems to select rivers in summer and slow flowing waterbodies in winter.[10] It does most of its feeding in the mornings and evenings and less during the day. Its diet has not been documented in detail. It is known to eat large crayfish and other river fishes like snow trout.

BREEDING

There is very little information on how this bird breeds. The bird is known to breed and roost in different species of broadleaved trees,[11] as well as in Chirpine trees. In Bhutan, until 2018 the nests of the birds were found only in Chirpine trees; however that year it was discovered that they also nest in broadleaved trees. In India, Dr Gopinathan Maheswaran, a scientist at the Zoological Survey of India in Kolkata and his research scholar Dr Himadri Sekhar Mondol, who now works as a scientist at the Bombay Natural History Society, were the first to discover the WBH's nests. They found this in Namdapha Tiger Reserve in 2014 and 2015 which proved that the species breeds in India. Nesting usually takes place in March–early June, and individuals then may move up to 30 km. to feed on sand bars in large braided river systems in winter.[12] Studies in Bhutan and India show that the WBH makes 'a simple platform nest' on tall trees some distance from water bodies with a clear overview of surroundings. They are sensitive to even slight disturbances.

PLATES 3.2a and 3.2b: Habitat of the White-bellied Heron
Image courtesy: Yumlam Benjamin Bida/ATREE

SOCIAL STRUCTURE

The WBH is generally a solitary species but is known to be found in small flocks and family groups during winter.[13] In early January 2014, six individuals were observed together[14] in Namdapha Tiger Reserve. However, mostly only one or two birds have been observed together.

CONSERVATION STATUS

The WBH is in the Critically Endangered[15] category of species, which is the highest grade of threat for a species given by the IUCN. In India it has also been upgraded to a Schedule I species from a Schedule IV species, thus giving it the same conservation status as other large mammal species such as tiger, elephant, etc. Other than in Bhutan, there haven't been systematic surveys in its potential habits or monitoring in its known habitats. We do not know how many birds still exist today in the wild. Globally the species is estimated to be less than 250 mature individuals. The numbers are continuing to decrease and no place in the world where it is found has more than 50 mature individuals. Scientists and conservationists are of the opinion that the species' numbers will continue to decline as more habitats are lost and degraded due to development-related activities.

THREATS

The WBH is threatened by multiple issues, most of which are related to anthropogenic activities:

Habitat degradation and loss: This includes widespread loss, degradation and disturbance around forests, water courses and wetlands. Except for Bhutan, in other countries where the species is present, we do not have a dedicated research and monitoring programme that collects information on the distribution, population and threats to the species. Some of the causes of habitat loss and degradation may be encroachment, overgrazing, conversion to agriculture and infrastructure development.

Poaching and destruction of nests: These are more local and direct threats reported to occur in its distribution range.[16] Poaching is thought to result in significant threats in key protected areas (e.g. the Namdapha Tiger Reserve) in north-east India and Myanmar while natural forest fires are known to destroy nests in Bhutan.

Hydroelectric dams: In Bhutan, hydroelectric dam developments as well as associated road improvements have resulted in significant habitat degradation, and the resultant affect on heron population.[17] Since the bird is dependent on fish for food, dams usually stagnate the river water and cover the banks. The birds is left with paucity of or no space to fish as the river bank gets inundated.

Disturbance: Riverbanks are used by rural communities to travel to different places. In the Namdapha Tiger Reserve one can often see people walking along these paths. During the tourist season there is an increased number of tourists, their guides and porters in most sites where the bird is found. The WBH is very sensitive to human disturbance. Scientists observed that it took two days for a pair of birds to return to their breeding site after it was disturbed. This causes much stress to the bird especially while feeding and during the breeding season.[18]

Stone and sand mining: Due to the proximity of villages to the species' habitats, activities like stone and sand mining were observed there. However, the intensity of this activity was nowhere in comparison to sand and boulder mining as well as stone crushing with huge machinery in the Noa-dehing river valley, which is the key habitat of the WBH. This can have a long-term impact on the species, especially due to the change in river architecture.

CONSERVATION ACTIONS

Several pertinent suggestions have been put forth by scientists, heron experts and conservationists for the conservation of the WBH:

Research programmes to understand the ecology and behaviour of the bird, with regard to especially breeding and population.

Conduct systematic surveys where it was present in the past, in areas where it is still present and also in sites that are potential habitats of the bird. This includes sites in West Bengal, Assam and Arunachal Pradesh.

Research programmes to study the ecology of the species in detail should be carried out based on two or three sites that can be chosen in the area where it is found.

Conduct genetic studies on the population structure, phylogeny, and distribution in different locations, extending across all range countries.

Identify key area specific threats and understand their motivation and causes from people's perspectives.

Sensitize, raise awareness and inform the general public, government, researchers, school and college students on the Critically Endangered status of the WBH and the various threats faced by the species.

NOTES

1. 'Species Factsheet: *Ardea insignis*', *BirdLife International* (website), 2020, see http://www.birdlife.org, accessed 2 June 2020.
2. M.R. Price and G.L. Goodman, 'White-bellied Heron (*Ardea insignis*): Conservation Strategy', IUCN Species Survival Commission White-bellied Heron Working Group, part of the IUCN SSC Heron Specialist Group, 2015.
3. G. Maheswaran, 'Foraging Ecology of the White-bellied Heron, *Ardea insignis*, in Namdapha Tiger Reserve, Arunachal Pradesh, India', *Progress Report to Science and Engineering Research Board*, (SERB), Department of Science and Technology (DST), Government of India, 2014; H.S. Mondal and G. Maheswaran, G., 'First Nesting Record of White-bellied Heron (*Ardea insignis*) in Namdapha Tiger Reserve, Arunachal Pradesh, India. Special Report', *Birding ASIA*, vol. 21, 2014, pp. 13–17.
4. T. Habung, K. Tasser, and T. Bamin, 'First record of White-bellied Heron *Ardea insignis* from Kamlang Tiger Reserve, Lohit district, Arunachal Pradesh, India', *Birding ASIA*, vol. 32, 2019, p. 120.

5. Gopinath Maheswaran, 'Walong's White Bellies', *Sanctuary Nature Foundation* (website), see https://sanctuarynaturefoundation.org/article/walong%E2%80%99s-white-bellies, accessed 25 January 2023; first published in *Sanctuary Asia*, vol. 41, no. 8, August 2021.
6. S. Ali, and S.D. Ripley, *Handbook of the Birds of India and Pakistan Together with Those of Nepal, Sikkim, Bhutan and Ceylon*, New Delhi: Oxford University Press, 1983; A.W. Tordoff, et al., 'Avifaunal Surveys in the Lowlands of Kachin State, Myanmar, 2003–2005', *Nat. Hist. Bull. Sim Soc*, vol. 55, no. 2, 2007, pp. 235–306.
7. B.F. King et al, 'Birds Recorded during Two Expeditions to North Myanmar (Burma)', *Forktail*, vol. 17, 2001, pp. 29–40.
8. BirdLife International, *Threatened Birds of Asia: The BirdLife International Red Data*, Cambridge: BirdLife International, 2001.
9. RSPN, 'The Critically Endangered WBH', ed. R. Pradhan, T. Norbu and P. Frederick, unpublished report, Thimphu: Royal Society for Protection of Nature, 2011.
10. BirdLife International, *Threatened Birds of Asia*.
11. I. Acharja, 'Evaluation of Nest Habitat, Site Preferences and Architecture of the Critically Endangered White-bellied Heron *Ardea insignis* in Bhutan', *Bird Conservation International*, 2019, pp. 1–19, see https://doi.org/10.1017/S095927091900042X.
12. R. Pradhan, *White-bellied Heron Project, 2005–2007*, unpublished Annual Report, December 2005–December 2006; R. Pradhan, T. Norbu, and P. Frederick, 'Reproduction and Ecology of the world's Rarest Arseid: The White-bellied Heron (*Ardea insignis*) in Bhutan', 31st Annual Meeting of the Water Bird Society, 30 October–3 November 2007.
13. Pradhan, *White-bellied Heron Project*.
14. Mondal and Maheswaran, 'First Nesting Record of White-bellied Heron (*Ardea insignis*) in Namdapha Tiger Reserve'.
15. 'Species Factsheet: *Ardea insignis*'.
16. Ibid.
17. Price and Goodman, 'White-bellied Heron'.
18. Mondal and Maheswaran, 'First Nesting Record of White-bellied Heron'.

4

Pangolin

'The Mammal You've Never Heard Of'

Sunita Pradhan

> Humans regard animals as worthy of protection only when they are on the verge of extinction.
>
> —Paul Craig Roberts

Tiger—the name itself evokes a familiar picture in anyone's mind, as the species is so popular and well known. On the contrary, a pangolin is a relatively lesser-known species. It is not charismatic and attractive as the tiger, red panda, or a snow leopard. Till about a decade ago, it was often referred to as 'mammal you've never heard of'.[1] Our interviews with children in pangolin habitats showed that many mistook the species to be a reptile when it is a mammal with scales and an awkward body with a long tapering tail.

However, pangolins are unique animals with amazing adaptations. There are animals like the South American anteaters and armadillos, which look very much like the pangolins. Despite these resemblances, pangolins come from very different ancestry and are, thus, very distinct from the armadillos or the anteaters. The pangolins are therefore classified in a separate order Pholidota,[2] all by itself under Class Mammalia.

PLATES 4.1: Chinese Pangolin (*Manis pentadactyla*)
Image courtesy: Kadoorie Farm & Botanic Garden

Fossil records of pangolins have been found in Europe, Africa, North America and Asia. The earliest fossil, Pholidota, dates back to the middle of the Eocene Epoch, some 56 million years ago in Germany.[3] Thus, although pangolins have a known evolutionary lineage in Europe, the eight currently found pangolin species in the world inhabit only the continents of Africa and Asia.

Out of the eight species of pangolin, four species range across Asia. The other four pangolin species are found in Africa—two Phataginus and two Smutsia.[4] The African Pangolins are the White-Bellied (*Phataginus tricuspis*), Giant (*Smutsia gigantea*), Ground (*Smutsia temminckii*) and Black-Bellied (*Phataginus tetradactyla*), while the Asian Pangolins are the Chinese (*Manis pentadactyla*), Sunda (*Manis javanica*), Palawan (*Manis culionensis*), and the Indian (*Manis crassicaudata*). India is home to the Chinese Pangolin and the Indian Pangolin.

Pangolins vary in size from 30 cm. to 100 cm. long and vary in weight from 1.50 kg. to 33 kg. according to species. The Giant Pangolin is the biggest pangolin while the White-bellied Pangolin is the smallest.

DISTRIBUTION

The eight extant pangolins are found in different places in Asia and Africa. The table below explains the extent of their distribution.[5]

Pangolin Species	*Distribution*
Indian Pangolin	Almost throughout India, Pakistan, Nepal, Bhutan, Bangladesh and Sri Lanka
Chinese Pangolin	Southern Nepal, north-east India, Bhutan, Bangladesh, Myanmar, northern Indochina, southern China including the island of Hainan, and most of Taiwan
Sunda Pangolin	Throughout East Asia including Brunei, Cambodia, Java, Sumatra, Borneo, Laos, Malaysia, Singapore, Thailand, Myanmar and Vietnam
Palawan Pangolin	Endemic to the province of Philippines
White-bellied Pangolin	Ranges from Guinea to Central Africa as far as south-western Kenya and north western Tanzania. Also found in Northern Angola and north western Zambia
African black-bellied Pangolin	Senegal, across the continent to Uganda, and south into Angola. They are also found in Congo Basin and Guinean forests.
Giant Pangolin	West Africa to Uganda, with the largest concentration in Uganda, Tanzania, and western Kenya.
Ground Pangolin	Southern and eastern Africa

PANGOLIN: THE SCALY ANTEATER

Typically, body hair and fur are the most important identifying characteristics of a mammal. However, it is not so with the pangolin. Although a mammal, the pangolin does not have visible hair or fur like other mammals. It is an armoured mammal. Its body is covered and protected with armour, like overlapping hard scales. These scales are made of a material known as keratin, which is a protein that makes up our nails and hair. The pangolin is toothless and has a diet predominantly consisting of ant and termites, captured using long and muscular tongues.[6] Its specialized diet of ants and its body covered with scales gives the pangolin, its common name, 'scaly anteater'.

PANGOLIN HABITS

The pangolin is mostly fossorial, which means the animal is adapted to digging and primarily living underground in burrows. A pangolin's strong claws and well-developed muscular system[7] help to dig the burrows. Burrows are an integral part of Pangolin's life and are talked about in a following section. All pangolins have short powerful legs. The hind limbs are longer and stouter than its forelimbs. All the limbs have five digits with strongly curved claws. The central three claws of the forelimbs are enlarged. All these physical features help the pangolin to dig the burrows it lives in, give birth, rear it young and find its food. However, some of the pangolin species, like the White-bellied Pangolin in Africa or the Sunda Pangolin from Asia, are arboreal[8]—they can climb and live on trees. To help and aid a pangolin to climb and live in trees, the species has developed a strong prehensile tail which has the capability to grasp things.

Pangolins have poor vision[9] but a developed olfactory system. Most pangolin species are nocturnal,[10] with their active time being in the night; they are only sporadically active during the day. As a pangolin is generally reclusive, timid and nocturnal, it proves difficult to sight and study them in the wild. Devices like camera traps and radio collars have been very useful in documenting activities of the pangolin.

A pangolin gives birth to usually one young after a pregnancy period of three to five months.[11] The young lives with the mother in the burrows, and does not venture out alone until they are about 3–4 months old. The mother carries the young pangolin on the base of its tail, when they are out of the burrow searching for food. As pangolins are hunted by people for meat and scales, we lose many pangolins every year in the wild, thus disturbing their population. Moreover, pangolins give birth to a single young in a year, which makes it difficult to recover their disturbed population, rendering them vulnerable.[12]

PANGOLIN'S DEFENCE MECHANISM

Alexandre Dumas, in his book *The Vicomte de Bragelonne* (1847), said that there are no creatures that walk the earth, not even those animals that have been labelled cowards, which will not show courage when required to defend themselves. Pangolins are timid animals, and do not have the capacity to attack or harm anyone but will definitely defend themselves from harm. A pangolin's defence mechanism is one of the most intriguing in the animal kingdom. The hard scales on its body are like armour, which provide protection against predators and can be used as a formidable weapon. The brown or relatively dark colours of its body scales allow a pangolin to blend in with various environments. However, on being threatened, a pangolin curls up into a hard, impregnable, tight ball. The muscles in the abdomen of the pangolin are strong and allow it to curl up and also prevent a predator from prying it open or attacking its midsection. Once curled into a ball, only the pangolin's sharp body scales are displayed whilst its soft belly is protected.

PANGOLIN HABITATS AND BURROWS

Depending on the species, pangolins occupy various kinds of habitats ranging from grasslands, primary and secondary forests, dry regions, bamboo forests, agricultural fields, plantations, or cultivated and fallow lands, basically supplying them with abundant food of social insects like termites and ants.[13]

PLATE 4.2: Rescued Chinese Pangolin being released in the wild
Image courtesy: Kadoorie Farm and Botanic Garden

PLATE 4.3: Wild Chinese Pangolin in Hong Kong
Image courtesy: Kadoorie Farm and Botanic Garden

The most important component of a pangolin habitat is its burrow. Pangolins dig burrows in areas relatively safe from disturbances. These burrows, in which they rest, are deeper and designed to provide the pangolin with a perfect shelter to stay, give birth and raise young ones. However, a pangolin can also dig burrows to search for ant mounds and termites. These burrows are known as feeding burrows. The feeding burrows are mostly found in clusters and are not as deep as the rest burrows.

PANGOLIN CONSERVATION STATUS

All eight species of pangolins are threatened and protected by national and international laws. However, the Chinese pangolin and the Sunda pangolin are listed as 'Critically Endangered' by the IUCN Red List of Threatened Species because of their drastic population decrease due to incessant exploitation of the species.[14] All eight species of pangolin are also listed in Appendix I of the Convention on International Trade in Endangered Species of Wild Fauna and Flora (CITES), banning international commercial trade.[15] In India, the Indian and the Chinese Pangolin are protected as an Endangered Schedule I species under the Indian Wildlife Protection Act (1972).

HOW IS THE PANGOLIN THREATENED?

The pangolin has now gained enormous recognition, greatly promoted by its large-scale illegal trade. Hunting and poaching for trade are the greatest threats to the survival of all pangolin species.[16] Pangolins are in very high demand as high commercial value foods and medicine,[17] both in Asia and Africa. A demand for pangolins, their derivatives, and products is reported from international markets such as the United States of America and other non-range states (Heinrich et al. 2016),[18] along with range states like China.[19] The pangolin is now widely recognized as an icon of illegal wildlife trade.[20]

Between 2000 to 2016, as many as 10 lakh pangolins were traded across the world. India also features as an important country where the pangolin is poached for trade.[21] Approximately 6,000 pangolins were found in the illegal wildlife trade in India during the period 2009–17, i.e. approximately 650 pangolins every year. Thus, the species is under severe threat.[22] Additionally, the pressure on pangolins is aggravated by rapid urbanization and use of pesticides in agriculture fields,[23] leading to potential loss of habitat.[24]

WHAT HAPPENS IF PANGOLINS GO EXTINCT?

The illegal wildlife trade has been catastrophic for pangolins and the ecosystems they inhabit. Overharvesting of pangolins threatens their existence. Any pangolin that is harvested could be a last remaining male or female, or could even be a pregnant female of that population. Such extractions could impact the structure of a population, leading to its extinction. Moreover, poaching and extraction disrupts the population of the species with an already slow reproductive rate. Apart from this, harvesting of pangolins also affects the food web and local ecosystems because pangolins strongly influence the abundance and diversity of important invertebrates such as ants and termites in these ecosystems.[25]

THINGS TO DO FOR PANGOLIN CONSERVATION

1. Stopping pangolin trade: It is very important to stop illegal wildlife trade of the pangolin species. The illegal trade of pangolin scales is largely driven by high demand and high profit.[26] Therefore, minimizing the demand of the pangolin and their body parts is important. Strategies to reduce the demand of pangolin products must be implemented so that the illegal trade decreases.[27]
   Wildlife trade also involves hunting by local people in the Pangolin habitat, and is a source of income.[28] Therefore, there is a strong need to look into other

earning opportunities for these people and support them by discouraging the pangolin trade and halt the rapid decline of the species.[29]

Although there are many good national and international laws to contain illegal trade and trafficking of pangolins, proper enforcement of these laws is very important to tackle the mass harvesting of this species in the wild.[30] Also, most of the illegal trade in pangolins involves multiple nations; hence proper coordination between different nations is a prerequisite to tack the problem of illegal wildlife trade. For this, enforcement agencies such as the police, customs, and other intelligence departments must work together to tackle this transnational crime.[31] The enforcement agencies must be very active and use various methods to combat the trafficking of ivory, rhino horns or tiger parts, etc.[32]

2. Protect habitats: Various types of land-use changes are damaging and encroaching on pangolin habitats. Judicious laws and monitoring must be in place to evaluate the impact of development and land-use change. Use of pesticides in agriculture and plantations must be controlled as this gravely harms the habitats of the burrowing pangolins.
3. Increase public awareness: Education and awareness building about the species, targeting its illegal trade, decreasing demand for use as medicine and food, decreasing habitat destruction, should be an important part of the pangolin conservation.
4. Research on pangolin species: We still know little about the current status and abundance of the pangolin species in their distribution ranges. Hence, there is a need for more research on the species.

Let the fight for saving this amazing 'mammal you've never heard of'—now an 'icon of wildlife trade'—not cease, but only intensify. May our future generations sight the pangolin, not just in books, but also in the wild.

NOTES

1. L.A. Harrington, N. Cruze, and D. Macdonald, 'Rise to Fame: Events, Media Activity and Public Interest in Pangolins and Pangolin Trade, 2005–2016', *Nature Conservation*, vol. 30, no. 107, 2018.
2. W.E. Zhang, Y.P. Ryder, and S.J. Brien, 'Molecular Phylogenetics and the Origins of Placental Mammals', *Nature*, vol. 409, no. 614, 2001, p. 6820.
3. T. Kormos, 'Manis hungrarica, n. sp., des erste Schuppentier aus dem europaischen Oberpliozan', *Folia Zoologica et Hydrobiologica*, vol. 6, 1934, pp. 87–94.
4. A. Hassanin, J.P. Hugot, and B.J. van Vuuren, 'Comparison of Mitochondrial Genome Sequences of Pangolins (Mammalia, Pholidota)', *Comptes rendus biologies*, vol. 338, no. 4, 2015, pp. 260–5.
5. The IUCN Red List of Threatened Species, 2020, see https://www.iucnredlist.org/, accessed 9 June 2020.
6. S.W. Choo et al., 'Pangolin Genomes and the Evolution of Mammalian Scales and Immunity', *Genome Research*, vol. 26, no. 10, 2016, pp. 1312–22.
7. Ibid.
8. Ibid.
9. D.A. Soewu and I.A. Ayodele, 'Utilisation of Pangolin (Manis sps) in Traditional Yorubic Medicine in Ijebu Province, Ogun State, Nigeria', *Journal of Ethnobiology and Ethnomedicine*, vol. 5, no. 1, 2009, p. 39.
10. D.W. Challender, C. Waterman, C., and J.E. Baillie, *Scaling up Pangolin Conservation: IUCN SSC, Pangolin Specialist Group Conservation Action Plan*, London: Zoological Society of London, 2014.
11. F. Zhang et al., 'A Note on Captive Breeding and Reproductive Parameters of the Chinese Pangolin, Manis pentadactyla Linnaeus, 1758', *Zoo Keys*, vol. 618, 2016, p. 219.
12. Ibid.
13. The IUCN Red List of Threatened Species, 2020, see https://www.iucnredlist.org/, accessed 9 June 2020.
14. Challender, Waterman and Baillie, *Scaling Up Pangolin Conservation*.
15. 'Appendices', *Convention on International Trade in Endangered Species of Wild Flora and Fauna* (website), 2017, see https://cites.org/eng/app/appendices.php, accessed 23 April 2019.
16. Challender, Waterman and Baillie, *Scaling Up Pangolin Conservation*; J.E.M. Baillie et al., 'Manis crassicaudata', The IUCN Red List of Threatened Species 2014 (website), 2016, see https://www.iucnredlist.org/species/12761/123583998, accessed 28 October 2016.

17. D.W.S. Challender, 'Asian Pangolins: Increasing Affluence Driving Hunting Pressure', *TRAFFIC Bulletin*, vol. 23, no. 3, 2011, pp. 92–3.
18. Sarah Heinrich et al., 'Where Did All the Pangolins Go? International CITES Trade in Pangolin Species', *Global Ecology and Conservation*, vol. 8, 2016, pp. 241–53.
19. Challender, 'Asian Pangolins'.
20. Zhang, 'A Note on Captive Breeding'.
21. Challender, Waterman and Baillie, *Scaling up Pangolin Conservation*.
22. Rajesh Kumar Mohapatra, 'Nearly 6,000 Pangolins in Illegal Wildlife Trade in India since 2009', *WWFIndia*, 2018, see https://www.wwfindia.org/about_wwf/enablers/traffic/media/?uNewsID=17061, accessed 7 November 2019.
23. Baillie, 'Manis crassicaudata'.
24. Challender, Waterman and Baillie, *Scaling Up Pangolin Conservation*.
25. J.M. Swart, P.R.K. Richardson, and J.W.H. Ferguson, 'Ecological Factors affecting the Feeding Behaviour of Pangolins (Manis temminckii)', *Journal of Zoology*, vol. 247, 1999, pp. 281–92; G.W. Sileshi et al., 'Integrating Ethno-ecological and Scientific Knowledge of Termites for Sustainable Termite Management and Human Welfare in Africa', *Ecology and Society*, vol. 14, no. 1, 2009.
26. Challender, Waterman and Baillie, *Scaling Up Pangolin Conservation*; Challender, 'Asian Pangolins'.
27. J.F. McEvoy et al, 'Two Sides of the Same Coin: Wildmeat Consumption and Illegal Wildlife Trade at the Crossroads of Asia', *Biological Conservation*, 2019, p. 238.
28. M. Rao et al, 'Hunting for a Living: Wildlife Trade, Rural Livelihoods and Declining Wildlife in the Hkakaborazi National Park, North Myanmar', *Environmental Management*, vol. 48, no. 1, 2011, pp. 158–67.
29. Harrington, Cruze and Macdonald, 'Rise to Fame'.
30. Ibid.
31. Zhang, Ryder, and Brien, 'Molecular Phylogenetics and the Origins of Placental Mammals'.
32. Kormos, 'Manis hungrarica'.

5

The Fall of Jatayu

The Oriental White-backed Vulture

Akash Deep Baruah

> Vultures are natural sanitary workers and they are absolutely essential for environmental and ecological balance.
>
> —MADRAS HIGH COURT, 2017

HOW LOW IT FELL!

ONE OF MY MOST endearing childhood memories is of flying kites in the campus of Assam Medical College in Dibrugarh, where my father was serving as a resident doctor. While flying the kite, I remember seeing vultures soaring high up in the sky, amidst the scattered wisps of cirrus clouds. No wonder the vulture is the highest-flying bird in the world! Some can even fly as high as 11 km. above the surface of the earth. As the years rolled by, I got busy with life and gradually lost touch with the sky. Destiny made me a forest officer and in the course of my service I started getting bad news about vultures. For reasons unknown, the vulture population in India had declined so alarmingly that by the year 2000, the species was almost on the verge of extinction. A similar drastic decline of vulture population was reported from Bangladesh and Nepal as well. Curiously, none of the usual reasons for decline of a wildlife species applied in the case of these birds. Vultures were neither hunted and nor were there any serious threats to their habitat. Carcasses, the primary food of the vulture, was not in short supply either. India, with the world's largest cattle population, offered an abundant source of carcasses to sustain a large population of vultures. Until the early 1990s, vultures were, in fact, the most numerous birds of prey in South Asia, including India. A study conducted across eighteen protected areas of the country from 1991–2 revealed that the population of vultures in India were in the tens of millions.

In Hindu mythology, Jatayu, the vulture king, had fallen from the sky after the demon king Ravana chopped off his wings. Jatayu gave up his life in his valiant effort to rescue Sita from the clutches of the evil king. However, in the real world, Jatayus were dying in thousands and no one had a clue which evil king was responsible for their unexplained deaths! Wildlife experts, foresters and environmentalists across the world were greatly puzzled at the sudden decline of vulture population. Of the nine species of vultures found in India, the population of the three species of *Gyps* vultures had plummeted from millions in the early 1990s to a mere 15,000 or so by mid-2000.[1] The Oriental White-backed Vulture (*Gyps bengalensis*) suffered the worst with nearly 99.90 per cent decline of its population between 1992 and 2007.[2] This bird, once the most numerous bird of prey in the country, has gone almost extinct in Andhra Pradesh, Tamil Nadu and Kerala and in parts of Rajasthan and J&K.[3] The population of the Long-billed Indian Vulture (*Gyps indicus*) and the Slender-billed Vulture (*Gyps tenuirostris*) had dropped by 97 per cent during the same period. In fact, the decline of vultures was so catastrophic that it earned the dubious reputation of registering the fastest decline of population for any wild species in the history of wildlife. The Passenger Pigeon, once the most numerous bird in the world, took

PLATE 5.1: An Oriental White-backed Vulture cleans a carcass to the bone. *Image courtesy*: Akash Deep Baruah

an even longer time before becoming extinct in 1914. In the year 2000, the IUCN conservation status of *Gyps* vultures dropped from 'Least Concern' to 'Critically Endangered' category. And this was just a step away from 'Extinct in Wild' status.

THE SEARCH FOR ANSWERS

It was Vibhu Prakash, the principal scientist at Bombay Natural History Society (BNHS), who first documented the sudden decline of vulture population in Rajasthan's famous Keoladeo Bird Sanctuary at Bharatpur. In 1988, he had counted 353 nesting pairs of vultures in the sanctuary. However, by 1996 their numbers had gone down by half and by 2000, not a single pair could be seen in the sanctuary. Something terrible had truly befallen these birds. To ascertain the reason of such deaths, an international team of scientists, under the aegis of the Peregrin Fund, began investigating the deaths of vultures in Pakistan in 2000. After three years of intense investigation, they finally discovered the reason for the decline of vulture population. The culprit was a drug called diclofenac sodium, a readily available non-steroidal anti-inflammatory drug (NSAID) used for relieving pain in livestock. Diclofenac is an effective and relatively inexpensive drug that was introduced in India in 1994 and was the drug of choice for veterinarians in the early 1990s. When vultures feed on carcasses of animals that have been treated with diclofenac up to seventy-two hours prior to their death, the toxification by diclofenac results in renal failure and eventual death of the birds. The dead birds exhibit clinical symptoms of visceral gout, i.e. formation of uric acid crystals within their tissues. This phenomenon of poisoning due to accidental consumption of contaminated meat is termed as unintentional or secondary poisoning. The finding was extremely significant and this led to a global protest against the use of diclofenac for treatment of livestock. A 2006 study on livestock carcasses in India had revealed that diclofenac was present in 11 per cent of the carcasses.[4]

Vultures are, however, not the only unfortunate raptors that have perished due to secondary poisoning. About half a century back, the Bald Eagle, the national symbol of the United States, and the Peregrine Falcon, the fastest bird in the world, were both pushed to the brink of extinction due to use of Dichlorodiphenyltrichloroethane (DDT), an organochlorine pesticide. DDT is a relatively inexpensive and effective pesticide that was used extensively in the United States. It was used initially to control malaria, typhoid and other insect-borne diseases. Later, it came to be widely used for pest control in food crops. However, DDT can persist in the environment for a long period and accumulates in the fatty tissues of wildlife, causing adverse health effects. As the Bald Eagle and Peregrine Falcons are top predators, they accumulated large amounts of DDT in their system through bio-magnification after eating birds that had fed on seeds, insects and fish contaminated with DDT. The result was diminished fertility, thinning of eggshells that resulted in crushing of the shells during incubation and unsuccessful hatching. The US Environmental Protection Agency banned the use of DDT in 1972 and since then the population of Peregrine Falcons and the Bald Eagles have risen in the United States. Incidentally, India is the only country still manufacturing and using DDT.

The California Condor (*Gymnogyps californianus*), a New World vulture, is North America's largest bird. It once ruled the skies from Washington state to Mexico and in Arizona and Utah. However, by the early 1980s, the condor became critically endangered due to loss of habitat, hunting and secondary lead poisoning. In 1982, only twenty-two condors were left in the wild. Being a scavenger, the bird often fed on carcasses of animals that have been hunted with lead bullets. Accumulation of lead in the condor's body resulted in anemia and organ failure, including brain damage, killing the bird in the process. In 2008, a ban was imposed against the use of lead ammunition for most hunting activities in California state's condor range. In 2013, California enacted a law barring use of lead ammunition throughout the state. Through an intensive captive-breeding program, more than 200 condors were later bred and released in the wild.

THE BATTLE FOR CONSERVATION

Following the discovery of secondary poisoning of vultures by diclofenac, the Ministry of Environment and Forests initiated India's Vulture Action plan in 2006, which provides strategies and actions for containing the decline of vulture population through ex-situ and in-situ conservation measures. The plan recommended banning veterinary use of the drug. The Government of India, as well as the governments of Pakistan and Nepal, banned the use of diclofenac in 2006, followed by Bangladesh in 2010. The Ministry of Health and Family Welfare, Government of India, through the notification G.S.R.No.499(E), dated 4 July 2008, published in the *Gazette of India* (Extraordinary) Part II, Section 3, Sub(Section) I, dated 5 July 2008, officially prohibited the use of diclofenac and its formulations for animal use. Prior to the ban, nearly seventy drug firms were producing this highly popular drug in India. However, despite the ban, human diclofenac formulations continued to be used illegally for veterinary purposes. In order to curb this practice, in 2015 the Government of India banned the production of injectable diclofenac products in vials larger than 3 ml. This made the drug inconvenient for use on livestock as they need larger quantities for a single dose. The normal veterinary dose for an adult cow was about 10 ml. (prescribed veterinary dose is 1000 mg./kg. of body weight). Two drug companies had challenged the government's decision in Madras High Court. However, the court in its landmark judgment of 24 October 2017 upheld the ban stating 'Vultures are natural sanitary workers and they are absolutely essential for environmental and ecological balance'.

Though diclofenac was banned, a number of other veterinary painkillers, such as ketoprofen, acelofenac, nimusulide, fluxinin, etc., were still available in the market that were toxic to vultures. Ketoprofen was found to be as lethal as diclofenac for vultures.[5] Further, acelofenac synthesizes to diclofenac in cattle and is, therefore, toxic to the vultures.[6] Studies also revealed that both nimusulide and fluxinin cause visceral gout in vultures.[7] Eventually, meloxicam was found to be a safe drug if ingested by vultures. Meloxicam is safe because it can be metabolized faster and, therefore, does not accumulate in the liver of the bird. It also has a shorter half-life, half an hour compared to twelve hours in case of diclofenac. Being a new-generation drug, meloxicam has fewer side effects compared to diclofenac. Currently, Melonex Plus, Melonex Power, etc., are some of the commonly prescribed Meloxicam formulations used for treatment of ungulates in India. Other than meloxicam, drugs such as analgin, fluxinin meglumine, etc., are also being used for treatment of livestock in the country. According to Dr Kushal Konwar Sarma, renowned veterinary surgeon and Padma Shree awardee from Assam, diclofenac is a very effective drug but unfortunately was maligned due to the death of vultures. Voveran, a diclofenac formulation is, however, widely used for treatment of pain and inflammation in humans.

In an effort to augment vulture population in the wild, eight ex-situ captive breeding centres have been set up in India; four in association with BNHS and four in association with the Central Zoo Authority (CZA). The Jatayu Breeding and Conservation Centre at Pinjore, a joint effort of BNHS and Haryana Forest Department, is the flagship vulture breeding centre in the country. The Vulture Conservation and Breeding Centre (VCBC) at Rani, set up in 2007 on the outskirts of Guwahati, is a joint initiative of the BNHS and Assam Forest Department. The centre has been breeding Slender-billed and Oriental White-backed vultures with funding from the Royal Society for Protection of Birds (RSPB) in the UK. According to Sachin Ranade, senior manager of the centre, captive breeding of vultures is a time consuming and challenging task due to a number of factors. First, the vulture is a slow-breeding species as it lays only one egg at a time and has an eight-month breeding cycle. Second, as vultures have a long lifespan of about forty to fifty years, they take about five to six years to reach adulthood. And before their release, the birds have to be trained to live in the wild in a pre-release enclosure. VCBC plans to release vultures after field studies confirm that diclofenac and other vulture toxic drugs are no longer a threat to the bird in the state of Assam.

Globally, secondary poisoning poses the greatest threat to the future of vulture revival. Other threats include habitat loss and degradation, decreasing food availability, human disturbance, collision with wind turbines and power lines. The Multi-species Action Plan to Conserve African-Eurasian Vultures (Vulture MsAP) was adopted in 2017 to provide a comprehensive strategic conservation action plan for conservation of fifteen migratory Old World vultures, the majority of which have been listed as 'Critically Endangered' by the IUCN. The multi-national initiative aims to promote a concerted, collaborative, and coordinated international action towards the recovery of the vulture population to acceptable levels by 2029. A blueprint for recovery of Asia's Critically Endangered Gyps Vultures was developed by the Saving Asia's Vultures from Extinction (SAVE) consortium. The said measures have found place in the Vulture MsAP. SAVE has also developed the concept of Vulture Safe Zones, i.e. creation of areas with a minimum 100 km. radius around important vulture breeding colonies that are free from use of harmful NSAIDs.

THE FALLOUT OF VULTURE DECLINE

While vultures were dying due to diclofenac, the species, with their unflattering looks and ungainly bearings, fails to evoke the needed sympathy in humans. Sadly, in our country, vultures are often associated with negative connotations such as a creature that relishes the prospect of someone dying, someone who lives off the dead, etc. The reality is, however, far from the truth. Vultures play a big role in cleaning the environment by feeding on carcasses that would otherwise be breeding grounds for harmful diseases and fatal pathogens such as anthrax, foot and mouth disease, etc. Vultures are very efficient scavengers and they can clean a large carcass to its bones in an hour or so. They are truly Nature's garbage collectors. Only vultures have the capability of neutralizing the harmful pathogens in carcasses due to the high acidity level of their stomach juices. Other animals such as rats, dogs and jackals that feed on carcasses can become carriers of harmful pathogens if they consume contaminated flesh.

Unknown to many, vultures have cultural values as well. In the past, the Parsi community did not cremate or bury their dead but used to leave the dead bodies in 'Towers of Silence' built on top of a hill or in a remote location to be consumed by vultures. The Parsi communities hold water and earth sacred and hence consider cremation or burying the dead as contributing towards polluting nature. This was practiced in the erstwhile Bombay Presidency area and in cities such as Bangalore, Hyderabad, and Kolkata. The decline of vulture population has now compelled the Parsi community to switch to using an electric crematorium. However, a similar practice of leaving dead bodies for consumption by vultures is followed by the Tibetans in Chinese-occupied Tibet. This practice, called 'sky burial', was possibly prompted by the lack of wood for burning bodies in the higher elevations of the Tibetan plateau. Further, the low oxygen level at high elevation makes efficient burning difficult. As the soil is mostly shallow in the rocky Tibetan plateau, finding places with suitable soil depth for burial can also be quite challenging.

The decline of the vulture population has resulted in several other adverse environmental consequences. A study in 2004 revealed that the decline in the vulture population had led to a sudden increase in the stray dog population in India, which went up from 18 million in 1987 to 25.50 million in 1997.[8] The decline in the vulture population resulted in increased availability of food for the stray dogs, leading to an exponential rise in their population. Studies indicate that the sudden rise in the stray dog population may have partially contributed to the rabies outbreak between 1992 and 2007 when 38.50 million additional dog bites were reported, resulting in the death of nearly 47,300 persons. Nearly 50 per cent of the victims were children below the age of 15.

EMERGING NEW THREATS

Although banning the use of diclofenac had neutralized the biggest threat to vultures, their plight is far from over. The remaining small populations are now facing a grave new threat, i.e. secondary poisoning by organophosphates. These are commonly used insecticides for agricultural purposes in India. In Assam, several incidents of secondary poisoning of vultures by organophosphates have been reported in the last few years from Sibsagar district. In April 2009, about twenty-one vultures, belonging to the *Gyps* species, died after consuming a poisoned goat carcass in Dikhowmukh area of the district. The carcass was reportedly poisoned as a retaliatory attack against a rabid dog which had killed a goat. On 21 April 2018, nineteen vultures, belonging mostly to the Slender-billed and White-backed species, died after they consumed the meat of a poisoned calf carcass in Sibsagar district. Again, in March 2019, a similar incident took place in Sibsagar district wherein thirty-four vultures died after consuming poisoned meat near Dolonipathar. In addition, in April 2020, nineteen vultures died in Disangmukh area due to pesticide poisoning after consuming poisoned carcass of a calf. Locals mostly cite poisoning as a retaliatory action against predators such as stray dogs and jackals that have killed their livestock. Some wildlife lovers are, however, skeptical of such reasoning. Though there have been instances of stray dogs killing small livestock, such activity has rarely been observed in the case of jackals, barring the occasional hunting of poultry. As jackals are scavengers as well, they are vulnerable to death through consumption of poisoned meat. Wildlife lovers suspect that there may have been deliberate attempts to hunt jackals for consumption of meat under the pretext of retaliatory action. The notion is not without merit but it needs further confirmation. According to Dr Kushal Konwar Sarma, there is a possibility that pesticide accumulation might be affecting calcium metabolism in vultures, leading to poor eggshell formation and crushing of the eggs during incubation. He, however, expressed that further studies needs to be carried out to confirm the same.

Apart from Sibsagar district, several instances of death of vultures due to secondary poisoning have been reported from other parts of the state as well. In March 2011, at least twenty vultures died in Goalpara district of Assam after consuming

PLATE 5.2: Vultures feeding on a carcass
Image courtesy: Akash Deep Baruah

a poisoned carcass. In February 2020, ten vultures died due to organophosphate poisoning in Chaygaon, near Guwahati. Poisoning of bait for large predators such as leopards and tigers has also resulted in the death of vultures in our country, particularly near protected areas. According to Sachin Ranade of VCBC, Rani, incidents of mass poisoning are a matter of serious concern in Assam. He had reported that in 2014 alone, 179 vultures were killed in seven separate incidents. Mr Ranade further informed us that around a hundred vultures die every year in the state due to numerous factors, including pesticide poisoning. VCBC at Rani has, till date, treated and released more than a hundred poisoned vultures to the wild.

All three species of vulture belonging to the *Gyps* family are listed under Schedule I of the Wildlife (Protection) Act, 1972. Hunting, including poisoning of vultures, can attract punishment up to seven years of imprisonment. In April 2019, one Ghanapati Das, a resident of Panitema village of

Kamrup district, was arrested by the Forest Department for causing the death of twenty-five Himalayan Griffon vultures by pesticide poisoning. On 1 June 2019, the Gauhati High Court granted bail to the accused on the condition that he would plant twenty-five saplings and take care of them until they grew big enough.

Decline of food availability for vultures is now a matter of growing concern. In a departure from the earlier practice of leaving carcasses in the open or in remote areas, people now resort to disposing off dead animals by burying them due to public resentment of leaving carcasses in the open. Furthermore, improvements in animal husbandry measures have also resulted in a lack of available carcasses for vultures.[9] This might affect the long-term survival of vulture population in South-East Asia, where diclofenac is not in use. To supplement the food supply for vultures, setting up of 'vulture restaurants' has been attempted at some places in India. A vulture restaurant is an undisturbed area where non-toxic, poison-free meat and carcasses are provided for

PLATE 5.3: Vulture 'restaurant' in Phansad Wildlife Sanctuary, Maharashtra. *Image courtesy*: Akash Deep Baruah

vultures and other scavengers. This supplementary feeding supports the vultures in times of food scarcity and when young birds fledge. In India, the first vulture restaurant was set up in Phansad Wildlife Sanctuary in Maharashtra. Besides food scarcity, habitat loss, forest degradation and reduced availability of suitable nesting trees for the vultures is also a matter of growing concern among wildlife lovers.

HOPE FOR THE FUTURE

In 2015, a nationwide road transect survey conducted by BNHS revealed that the population of the Oriental White-backed Vulture was around 6,000 in India,[10] of which 1,000 are possibly found in the state of Assam. Global population of this species was unlikely to cross 8,000. The same study estimated the population of the Long-billed and Slender-billed Vulture at around 12,000 and 1,000, respectively. Although this is a huge drop from the millions in which they had once numbered, it is encouraging to learn that the population of the Oriental White-backed Vulture is on the rise, as reported in a survey published in the *Cambridge Journal* in December 2017. Sadly, the population of the Indian Vulture still continues to decline. The status of the Slender-billed Vulture is presently difficult to predict due to its small population size. The 2015 study, however, indicated that the rate of decline of this species in India has slowed down and that the population might be stable. Populations of the vulture have also risen in the southern states of Tamil Nadu and Kerala in the past several years. Though diclofenac is no longer in use in our country, it will take some time for a slow breeding bird like the vulture to soar the skies once more in their earlier numbers. Recent conservation efforts in the country and in different parts of the world augurs well for the future of the vulture. And, as the title of a popular James Hadley Chase thriller *The Vulture is a Patient Bird* reads, the bird shall surely wait for better times to rule the skies once again.

NOTES

1. Vibhu Prakash et al., 'Recent Changes in Populations of Resident Gyps Vultures in India', *Hist. Soc. Journal of the Bombay Natural History Society*, vol. 104, 2007, pp. 129–35, see https://www.researchgate.net/publication/228495035_Recent_changes_in_populations_of_resident_Gyps_vultures_in_India, accessed 20 December 2022.
2. Ibid.
3. 'Multi-species Action Plan to Conserve African-Eurasian Vultures', IUCN (website), see https://portals.iucn.org/library/sites/library/files/documents/2017-064.pdf, accessed 20 December 2022.
4. M.A. Taggart, 'Analysis of Nine NSAIDs in Ungulate Tissues Available to Critically Endangered Vultures in India', *Environmental Science & Technology*, vol. 43, no. 12, 2009, pp. 4561–6.
5. Ibid.; Naidoo Vinny et al., 'Toxicity of Non-Steroidal Anti-Inflammatory Drugs to *Gyps* Vultures: A New Threat from Ketoprofen', *Biology Letters*, vol. 6, no. 3, 2009, pp. 338–42.
6. T.H. Galligan et al., 'Metabolism of Aceclofenac in Cattle to Vulture-killing Diclofenac', *Conservation Biology: The Journal of the Society for Conservation Biology*, vol. 30, no. 5, 2016, pp. 1122–7.
7. Richard Cuthbert et al., 'Continuing mortality of vultures in India associated with illegal veterinary use of diclofenac and a potential threat from nimesulide', *Oryx*, vol. 50, no. 1, 2016.
8. Alok Gupta, 'Carcasses in the Ganga', *The Wire*, April 2016, see https://thewire.in/environment/carcasses-in-the-ganga, accessed 20 December 2022.
9. Vibhu Prakash et al., 'Recent Changes in Populations of Critically Endangered Gyps Vultures in India', *Bird Conservation International*, vol. 29, 2017, pp. 1–16.
10. Ibid.

6

Great Indian Bustard The Vanishing Prince of the Grassland Kingdom

Yogendra Singh Rathore

> The natural world is in serious trouble and it needs our help as never before. We continue to lose the precious wildlife that enriches our lives.
>
> —David Attenborough

The great Indian bustard (GIB) or Godawan (*Ardeotis nigriceps*), a majestic species among twenty-six globally known family members of bustards, is facing a severe threat of disappearance from this planet. The magnificent bird was once a strong candidate for the representation of our national pride (as national bird) and was strongly supported by the Indian Ornithologist Dr Salim Ali, but disregarded due to its name, which might be misspelt with 'Bastard'. Eventually, the Indian Peafowl (*Pavo cristatus*) was crowned with the 'National Bird of India' tag which is apparently less susceptible to mutilation. Ignorance remained blissful for the governments (both Centre and State) until the GIB reached the brink of extinction and was listed as a 'Critically

Endangered' species in the IUCN Red List of Threatened Species. It got international recognition only when a handful of individuals were left in the wild, serving as a 'wake-up call'. Currently, we have less than 150 birds on this entire planet which are the last hope to recover this species to a healthy population level.

GENERAL DESCRIPTION OF THE SPECIES

Globally, there are twelve genera (*Heterotetrax* being split out from *Eupodotis*) and twenty-six different species of bustards in family *Otididae* with several curious biological attributes. The family members lack hind toes (and therefore cannot roost on trees), rarely fly, and have a strong terrestrial and vagrant character.[1] Since the majority of bustards (twenty-one of twenty-six extant species) are mainly distributed in Africa, it is expected that the lineage originated there some 30–50 million years ago.[2] Some other report suggests that bustards originated 77 million years ago in Africa and speciated over discreet ranges of the Old World grasslands.[3] The GIB was initially identified and named *Otis nigriceps*[4] after a scientific analysis of specimens collected at the Himalayan foothills of north-west India. Later on, the name was chronologically changed to *Otis edwardsi*,[5] *Eupodotis edwardsi*,[6] *Choriotis nigriceps*[7] and lastly *Ardeotis nigriceps*.[8]

GIB is very selective and habitat-specific bird, and if it does not find a habitat suitable, it straightaway deserts that particular place and even its immature eggs. As GIB is an obligate grassland species, it is considered the 'flagship' species of grassland ecosystems due to its habitat association, charismatic nature and extinction-prone, slow life history traits. It is highly reliant on traditional farming and traditional practices of grassland management. Savannah/scrublands, open country, short grasslands and agro-grass mosaics are the typical GIB habitats. The dry savannah grasslands of the hot arid and semi-arid regions of peninsular India with average rainfall, varying from 400 to 1,000 mm., are considered prime and pristine habitat for GIBs.

In past, the GIB has been well known as a game bird and being hunted for many years. In 1966, the species was described as 'very rare and apparently decreasing' and therefore informally listed as threatened for the first time.

CLASSIFICATION/TAXONOMY

Kingdom	*Animalia*
Phylum	Chordata
Class	Aves
Order	Otidiformes
Family	Otididae
Genus	*Ardeotis*
Species	*Ardeotis nigriceps*
Synonyms	*Choriotis nigriceps*

MORPHOLOGY AND BEHAVIOUR

The GIB is a diurnal bird species which remains active during early morning (6.00–11.00 a.m.) and evening (5.00–8.00 p.m.) hours. It prefers to roost in open bare grounds during nights while during the day a patch of thick grass or the shade of a bush is used for roosting. It has a unique flocking behaviour where both the male and female bustards live in their own flocks (sexually segregated) and mixed flocks are either rare or absent.[9] They may form a temporary mixed flock for foraging and roosting in areas where food is in abundance. The flocking appears more prominent during roosting behaviour. The average size of a flock also varies season to season. The flock size generally reaches its maximum just before the onset of breeding season and remains minimal during the breeding season.

PHYSICAL FEATURES

The GIB is a magnificent, hefty monotypic species which is highly endemic to the open agro-grasslands of the Indian subcontinent with surviving population in India and Pakistan only. In general, the GIB is a silent bird, but when threatened it produces barking sounds. It flies very swiftly with neck extended and head low. It is a tall, long-necked, and long-legged bird (similar to the ostrich) and highly adapted to its cursorial (adapted specifically to run) nature. Morphologically, the GIB male is generally larger than the female. The male GIB measures, 1.10–1.20 m. in length and weighs 8,000–14,500 gm. The female is smaller and measures 70–90 cm. in length and weighs 3,500–7,000 gm. Morphometric data exhibits that the foot size of the male GIB is larger than the female.[10] The length to width ratio between male and female bustard does not have a significant difference.

REPRODUCTION OR BREEDING BEHAVIOUR

The GIB is a gregarious/semi-gregarious bird and shows sexual dimorphism with distinct secondary sexual characteristics. Sexual maturity is attained by GIB at the age of 3–4 years (female) and 4–5 years (male), and it may reproduce up to an age of 20 years.[11] The species is capable of year-round breeding, but most of the populations breed principally between March and September (largely, the summer monsoon season) with some rainfall-based regional variations due to availability of adequate food and shelter. The male develops a black breast band during the breeding season and typically is a deep sandy buff body colour. During courtship, the males' crown is crested and puffed up and they produce deep, resonant booming mating calls to attract breeding females.

The female lays only one clutch per year which consists of one (and very rarely two) pale olive-brown eggs during the breeding season. The eggs are laid on open ground with slight elevation from the plain giving the incubating mother a clear line of sight over the landscape. The eggs are mostly laid in the month of May and onwards.[12] The egg is incubated

A

B

PLATES 6.1a and 6.1b: Great Indian Bustard male (A) and female (B) walking in open habitat. A typical black breast band is clearly visible in breeding male. *Image courtesy*: Dr Divesh Saini

A

B

PLATES 6.2a and 6.2b: Male GIB in its typical habitat showing courtship behaviour characterized by cocked-up tail and inflated gular pouch
Image courtesy: © Dr G.S. Bhardwaj, IFS

PLATE 6.3: The GIB's egg collected for artificial incubation in lab
Image courtesy: © WII Bustard Recovery programme

by the female for about 25–7 days without any support from the male (as they are polygamous) in nest safeguarding. After hatching, the newborn chick fledges in about seventy-five days and remains dependent on the mother for food and protection. The chick predominantly feeds on insects like grasshoppers, beetles and small lizards. When threatened, female birds carry dependent young chicks under their wings and can indulge in confrontation with even larger animals like foxes to save their chicks.

The GIB is a k-selected species, which means that it produces a lower number of offspring and invests significant energy in rearing one offspring at a time. The offspring takes a longer period of maturation to adulthood and need heavy parental care and nurturing with aggressive protection of the juvenile by the parents. Life expectancy is higher in K-selected species[13] and it follows the Type I or Type II survivorship curve.[14]

FEEDING BEHAVIOUR

The GIB is an opportunistic omnivore and in the wild their diet reflects the local and seasonal abundance of plants and animals.[15] According to the seasonal shifting of diet it remains predominantly herbivorous in winter and insectivorous in the summer and monsoon. It generally feeds on: (1) seeds of grasses, berries of *Zizyphus* and other plant matter; (2) insects and arthropods like grasshoppers, beetles, crickets, mole crickets, mantis, water termites, larger ants, caterpillars, centipedes, spiders, worms, scorpion and locusts; (3) vertebrates like rats and lizards; and (4) agricultural crops such as exposed groundnuts, millets, pods of legumes and wheat grains. Being a desert fauna, it is a facultative water drinker and drinks water regularly only if available but can survive for a longer time without it.

DISTRIBUTION RANGE

Generally, the ecological occurrence of a species over a geographical area is termed as distribution range. The information about distribution of a species is very important in order to conduct ecological research. Various aspects of the distribution range and pattern help in formulating robust conservation management decisions for the species under study. The GIB was formerly widespread in the plains of the Indian peninsula, particularly in the grassland ecosystems of India and Pakistan. W.T. Blanford mentions the distribution of the GIB in:

> the plains of the Punjab between the Indus and the Jumna, also Eastern Sindh, Cutch, Kattiaywar, Rajputana, Guzerat, the Bombay deccan, the greater part of the central provinces, extending as far as east Sambalpur, the Hyderabad territories, parts of the Madras presidency and the Mysore state as far south as southern Mysore and perhaps further south.[16]

In India, historical distribution of the GIB was throughout the western half of the country, from central Punjab in the far north to central Tamil Nadu in the south, and from western Orissa in the east to the western borders of Gujarat and Rajasthan, spanning eleven Indian states.[17] The last three decades have witnessed a drastic decline in the range occupancy of the GIB, where the species is reportedly extinct from 90 per cent of its former range.[18] Its current range occupancy is severely confined to fragmented patches of Rajasthan, Gujarat, Maharashtra, Karnataka, Madhya Pradesh and Andhra Pradesh in India, and Sind in Pakistan, in scattered and isolated populations.[19] It has practically vanished from Punjab, Haryana, Uttar Pradesh, Orissa and Tamil Nadu except for vagrant individuals here and there. Some may still survive in Cholistan and Sind in Pakistan but the largest population is thriving in continuous stretches of western Rajasthan.[20] In Rajasthan, a dedicated survey carried out by Dr A. Rahmani and his team revealed that the bustard is found in eleven districts of Rajasthan state, namely Jaisalmer, Jodhpur, Bikaner, Ajmer, Kota, Tonk, Sawai Madhopur, Barmer, Pali, Jalore and Bhilwara, with the first five districts having confirmed sightings. At present, the highest surviving population of the GIB is nearly confined to the Thar Desert in Jaisalmer and Bikaner districts.

HABITAT PREFERENCE

The GIB is an elegant bird species found in grassland ecosystems. It favours arid and semi-arid areas dominated by grasslands with flat or gently undulating topography containing scattered low thorn scrub, bushes and tall grass interspersed with low intensity cultivation for nesting, roosting, foraging and display.[21] Low erratic rain, droughts, scattered cultivable dry lands with sparsely distributed water sources are the prime characteristics of its habitat. The habitat of this bird species is not uniform due to a large distribution range and therefore in some parts of its range it may occupy entirely dry habitat. Additionally, its habitat requirements may also vary for different activities according to season and behaviour.

The five general habitats required by this species are:

1. Breeding and nesting habitat: This comprises of open flat lands with short to medium length grass (up to 50 cm.)

patches which are primarily characterized by a mosaic of sparsely grazed tall grass with minimal outsider disturbance.[22] During summer and the monsoon season, they congregate in such traditional undisturbed grassland patches. They are very sensitive about their nesting and breeding habitats and are known to abandon them if observe any disturbance.

2. Display habitat (Lekking sites): Display habitats are primarily used by displaying male bustards, which involves open plain ground with slightly elevated sites and sparse herbaceous vegetation (< 25 cm.) to provide better visibility.[23] These are traditional sites where males gather to display and attract females.
3. Foraging/Feeding habitat: The feeding habitats are composed of a mix of various habitats like scrublands, grassland, pasture, crop fields and stony wasteland with plenty of food resources. They favour short grass cover which improves the foraging success by increasing their visibility and improving mobility to catch their prey.
4. Roosting habitat: The GIB requires moderate shade (with 25–50 cm. vegetation cover and small trees or shrubs) in the day and bare ground with minimal scrubs at night for roosting.[24] It may also require little taller grass (~75 cm.) for resting and chick-rearing.
5. Wintering habitat: Apart from these above stated habitats, the GIB remains vagrant in wide and sparse natural agro-grass-scrub landscapes in the non-breeding season and probably makes local and long-distance nomadic movements under the influence of various factors.[25] Such habitats act as wintering grounds for non-breeding bustards.

THREATS

The number of this majestic bird has dwindled very fast in the recent past. Being an obligate grassland taxa and due to their predominantly slow terrestrial life history and ground nesting habits,[26] the GIB is vulnerable to many threats. As a result, the GIB population is continuously falling, from around 1,260 in 1969 to 300 in 2008; only around 150 are left at this moment.[27] Detailed understanding of threats is very important to understand the trajectory of species loss and to develop conservation and management strategies within time. The threats which are responsible for the continuous decline of GIB population are as following:

1. Habitat destruction and degradation, fragmentation of habitat and habitat alteration.
2. Indiscriminate hunting and poaching for alleged aphrodisiac value and occasional egg collection for consumption (most incidents are in Pakistan, but low-intensity poaching still persists within India as well).
3. Overgrazing of grasslands due to lack of well-defined grazing policy.
4. Intensive use of insecticides and pesticides by farmers to control pests.
5. Anthropogenic infringement in the fragile landscape.
6. Power-lines, particularly high-voltage (33–440 KV) transmission lines with vertical alignment responsible for electrocution of flying bustards. High-voltage transmission lines across the GIB habitat cause unsustainably high mortality of about 15 per cent of the GIB population in the Thar region alone.
7. Ongoing developmental works of pipeline and canal digging.
8. Unsustainable and intensive agriculture, shifting of agricultural preferences from GIB-friendly (sorghum, millets, groundnuts, pigeon pea, etc.) to more profit-making crop production (cash crops like sugarcane, grapes, cotton and horticulture, etc.) and changing cropping practices from seasonal to year-round, intensive crops.
9. Drastic hydraulic changes and massive agricultural conversion due to the Indira Gandhi Nahar Project.

10. Increasing industrial pressure on land along with their associated infrastructural developments.
11. Anthropogenic disturbances including urbanization, human encroachments, disturbances from local farmer's and daily farm worker activities, and illegal construction activities.
12. Stone quarrying, mining and its associated activities which creates an imbalance in the ecological equilibrium.
13. Natural predators like wolves, jungle crows, foxes, jackals, monitor lizards and mongoose which targets eggs and chicks of the GIB. The natural success rate of GIB eggs is about 60–70 per cent. which is further lowered to 40–50 per cent due to these predators.
14. Disturbance of breeding and incubating bustards by stray dogs.
15. Disturbance due to transportation (rapid expansion of linear developments) and service corridors and their associated infrastructures.
16. Ill-informed forestry practices by state governments such as plantation of exotic shrub/tree species (Eucalyptus, *Glyricidia* and invasive *Prosopis juliflora*) in deserts and grasslands in the name of afforestation.
17. Disturbance by freely grazing and semi-nomadic livestock and associated ungulate species which can reach very close to the nest, trample the eggs and disturb the incubating bustard.
18. Renewable energy production activities like wind turbines and solar energy projects near the core habitat of the GIB.
19. Uncontrolled overgrowth of invasive species in the habitat.
20. Proposed and ongoing thermal power projects and coal mines, and increasing atmospheric temperature due to existing thermal power projects.
21. Mismanagement and negligent behaviour of responsible agencies.

22. Disease and parasitic infections.
23. Disturbance due to noise pollution which affects mating and courtship practices of the GIB.
24. Disturbance due to unethical nature photography during breeding season.
25. Lack of legislative enforcement and protective measures in core breeding areas.
26. Lack of awareness among locals about breeding of the GIB and lack of public support for conservation initiatives. Severe antipathy of the locals towards GIB conservation and its associated landscape/sanctuaries.

CONSERVATION STATUS AND CONSERVATION ACTION (PAST AND PRESENT)

The GIB was one of the most popular game birds and therefore hunted for sport and food since the time of British rule in India. It has continued to be hunted in neighbouring Pakistan.[28] Continuous hunting and habitat destruction resulted in a drastic decline in numbers from about the middle of the twentieth century. After Independence, India emphasized the protection of GIB, citing its rapidly dwindling number. In 1952, the GIB was included in the fully protected list of endangered birds by the Indian Board of Wildlife (now known as National Board of Wildlife). All the states accepted the recommendations provided by the IBWL and initiated conservation of the GIB by setting up protected areas in its distribution range. However, the number was still plummeting sharply and therefore the preservation of the GIB was re-emphasized at the IUCN general assembly held at New Delhi in 1969. Unfortunately, hunting of the GIB along with Houbara Bustard and Lesser Floricans by royals and state guests continued till 1978 in Rajasthan. It is pertinent to mention here that 1978 was a watershed year in the history of GIB conservation. At the end of this year, a high-level campaign was started against government-supported illegal falconry by the Arab sheikhs, under the leadership of Mr Harsh Vardhan of Jaipur, Rajasthan. This was one of the most successful 'Public Movements' across the globe and was supported by many conservation organizations like the Bombay Natural History Society (BNHS) and the World Wildlife Fund (WWF India). The Jodhpur High Court issued a stay order over illegal falconry in the Thar Desert. Finally in 1979, the GIB was accorded Schedule I protection in the Indian Wildlife (Protection) Act 1972 and hunting of the GIB was completely banned by law with provisions of severe legal prosecution. Later on, the campaign resulted in an international level meet, the International Symposium on Bustards (ISOB), which was held in Jaipur in November 1980.[29] A 'postal stamp' depicting a male 'Great Indian Bustard' with a female in its typical habitat was issued at this landmark symposium.

PLATE 6.4: Postal stamp of the Great Indian Bustard dated 1 November 1980. *Image courtesy*: http://www.indianpost.com/viewstamp.php/Alpha/G/GREAT%20INDIAN%20BUSTARD)

Since 1981, extensive fieldwork has been done to understand the status, distribution and ecology of the GIB and its conservation status has been kept updated over time.

TABLE 6.1: Red List Assessments of Great Indian Bustard till Date

Sr. No.	*Year of Assessment*	*Red List Category*	*Approximate No. of Individuals**
1.	1988	Threatened (T)	1,260
2.	1994	Endangered (EN)	745
3.	1996	Endangered (EN)	-
4.	2000	Endangered (EN)	600
5.	2004	Endangered (EN)	-
6.	2008	Endangered (EN)	300–50
7.	2011	Critically Endangered (CR)	150–75
8.	2012	Critically Endangered (CR)	-
9.	2013	Critically Endangered (CR)	-
10.	2015	Critically Endangered (CR)	-
11.	2016	Critically Endangered (CR)	-
12.	2017	Critically Endangered (CR)	-
13.	2017	Critically Endangered (CR)	-
14.	2018	Critically Endangered (CR)	≤150

*Data may vary based on different sources of reporting

Recently, the GIB has also been included in Appendix I (species threatened with extinction) of the UN Convention on Migratory Species (UN CMS or Bonn Convention) at the 13th Conference of the Parties (COP) held in Gandhinagar (Gujarat) in February 2020. The GIB was projected as the 'mascot' of CMS COP13. India is a signatory country of both CMS and CITES conventions and hence it is under an international obligation to protect this species.

In 2009, the GIB (including two other bustard species) was identified as one of the priority species under the

'Recovery programme for saving critically endangered species and habitats' of the Centrally sponsored scheme 'Integrated Development of Wildlife Habitat' (IDWH), wherein financial assistance is given for said species. The Bustard Task Force conceptualized the 'Guidelines for preparation of State Action Plan for Resident Bustard's Recovery Programme' through various consultative meetings with forest Departments, wildlife biologists, non-governmental conservation agencies and other stakeholders. A paradigm shift was observed in the conservation methodology of the GIB, from protecting only species (orthodox model of conservation) to conserving the species and associated landscape (holistic conservation and management approach) which was missing from the earlier plan. Although the GIB was enjoying legal protection under Acts and other conventions, its habitat was still considered 'wastelands' and was not regulated by laws. Therefore, under this approach, conservation efforts primarily focused on the maintenance and management of such landscapes that showed long-term site fidelity by bustards. Rajasthan became the first state to initiate 'Project Great Indian Bustard' on World Environment Day 2013, under the proposed guidelines of the Species Recovery Programme. Currently, the Thar region of Rajasthan holds more than 75 per cent of the total GIB population across the globe and, therefore, the main focus is on Rajasthan. In Rajasthan, Desert National Park (actually a wildlife sanctuary) is the prime breeding and non-breeding ground for the GIB.

Since the 1980s, conservationists have been eagerly discussing a plan of action to protect this vanishing species and many state governments participated in such projects. However, one approach that is of the utmost necessity is Species Recovery Programme (SRP). SRP emphasizes the holistic conservation approach of species and habitat levels to secure long-term persistence of the resident bustard population of India. The SRP recommends:

(A) protection of core areas by making them inviolate of consumptive human use during breeding months (June–September); (B) landscape management plans that integrates livelihood concerns with conservation by encouraging compatible livelihoods and curtailing harmful infrastructure in priority areas; (C) population research and monitoring, and habitat research using biotelemetry across the range; (D) conservation breeding as a security policy and (E) a liaison committee between Government and local communities for joint decision making on incentive driven pro-conservation use of private-public mixed ownership lands.[30]

Keeping these guidelines in mind, the necessary recovery actions are divided into two broad sub-parts:

1. Species Level Recovery: (a) As the species remains vagrant without any specific boundaries, a systemic and coordinated scientific study of population ecology at the national and state levels is required using modern techniques and theories. (b) In view of the extremely low number of GIBs in the wild, there is an urgent need for ex-situ conservation strategies like captive breeding programmes to supplement the rapidly vanishing in-situ population. The captive stock may be treated as an insurance against total extinction and can be reintroduced later in the wild after successful restoration of habitat.
2. Habitat Level Recovery: (a) The prioritized bustard areas need to be freed from all kind of anthropogenic disturbances (excluding low intensity traditional agro-pastoralism) and the breeding areas must be consolidated through chain-link predator-proof fencing. (b) Integrated farmer-bustard friendly and landscape-dependent conservation policies need to be implemented through community sensitization. Environmentally viable grazing and land tenure policy in bustard landscapes is required for effective management of bustard habitats. Ill-informed forestry practices need to be curbed and multi-departmental coordination with accountability should be promoted. (c) Suitable

> breeding and non-breeding bustard habitat must be clearly identified through research and these must be monitored over the years to know seasonal critical resource requirements and landscape scale distribution pattern.

The Wildlife Institute of India (WII), the Ministry of Environment, Forest and Climate Change of India (MoEFCC) and state governments are collectively working towards ex-situ and in-situ conservation of the GIB. In 2018, an action plan of collecting twenty-two to twenty-four eggs in three years was prepared to create a founder population. An artificial hatching centre was set up the Sam region of Jaisalmer, where the eggs have been incubated under controlled conditions. The GIB breeding centre was established in 2019 and, since then, twenty chicks have hatched successfully. A predator-proof enclosure of about 18 km. in circumference (15.10 sq. km.) has been constructed to facilitate egg laying without any threats or human disturbances.

Recently, scientists working on ex-situ breeding projects have observed an altogether new trend of female GIBs laying two eggs in a clutch. This behaviour is extremely rare in the GIB and possibly this observation from DNP, Jaisalmer is the first report of GIB producing two-egg clutches with such a high incidence. In 2020, two GIBs were reported to lay two-egg clutches, and the same phenomenon has been repeated by another four GIB females in 2022. Observations suggest that the GIBs have demonstrated this new habit after consuming an additional protein-rich diet during these two years in addition to the conservation efforts taken locally. In 2020, there had been a locust attack in the region which provided protein-rich feed and in 2022, good rainfall resulted in luxuriant growth of *sewan* grass (*Lasiurus scindicus*) which in turn promoted a number of insects and pests in the local moist and humid environment.

Such reports are definitely a matter of happiness and can pave the way to enhance positive conservational interventions.

PLATE 6.5: Aerial view of the GIB captive breeding centre/artificial hatchery in Sam, Jaisalmer. *Image courtesy*: Sajal Sharma, WII

After analysing the breeding results of the last three years, construction of one more such enclosure had been proposed in DNP for which approval has also been granted. The WII team working on ex-situ breeding of GIB is planning to collect only one of the two eggs from the two-egg clutch for artificial hatching while the second egg will be reared by the bird itself.

THE CLARION CALL

Extinction of a species is a catastrophic event and India is not ready to lose another species, which has already happened with the Indian Cheetah (*Acinonyx jubatus*) in the post-Independence era. In this case, we are already late in taking appropriate timely actions and this is the last chance to prevent the inevitable extinction of the GIB in a true sense. India has a rich tradition of biodiversity conservation and it is integrated in the cultural ethos of our people. Therefore, it will be a great sense of dishonour and disappointment internationally if we fail to save this majestic prince of the grassland kingdom.

NOTES

1. J. del Hoyo, A. Elliott, and J. Sargatal, *Handbook of the Birds of the World: Hoatzin to Auks*, vol. III, Barcelona: Lynx Edicions, 1996.
2. O. Broders, T. Osborne, and M.A. Wink, 'Mtdna Phylogeny of Bustards (Family Otididae) Based on Nucleotide Sequences of the Cytochromeb-gene', *Journal für Ornithologie*, vol. 144, 2003, pp. 176–85; C. Pitra et al., 'Phylogenetic Relationships and Ancestral Areas of the Bustards (Gruiformes: Otididae), Inferred from Mitochondrial DNA and Nuclear Intron Sequences', *Molecular phylogenetics and evolution*, vol. 23, 2002, pp. 63–74.
3. Hoyo, Elliott and Sargatal, *Handbook of the Birds of the World*; P.A. Johnsgard, *Bustards, hemipodes, and sandgrouse: birds of dry places*, New York: Oxford University Press, 1991; S. Dutta, A.R. Rahmani, and Y.V. Jhala, 'Running out of Time? The Great Indian Bustard Ardeotis Nigriceps—Status, Viability, and Conservation Strategies', *European Journal of Wildlife Research*, vol. 57, 2010, pp. 615–25.
4. Vigors, 'Great Indian Bustard (*Ardeotis nigriceps*)', *Proceedings of the Zoological Society of London*, 1830–1.
5. J.E. Gray, *Illustrations of Indian Zoology*, London: Treuttel, Wurtz, and Richter, 1831.
6. W.T. Blanford, *The Fauna of British India, including Ceylon and Burma*, vol. III, London: Taylor and Francis, 1895.
7. S. Ali and S.D. Ripley, *Handbook of the Birds of India and Pakistan*, vol. II, Delhi: Oxford University Press, 1969.
8. C.G. Sibley and J.E. Ahlquist, *Phylogeny and Classification of Birds: A Study in Molecular Evolution*, New Haven and London: Yale University Press, 1990.
9. A.R. Rahmani, 'Flocking Behaviour of a Resident Population of the Great Indian Bustard Ardeotis nigriceps (Vigors)', *Revue d'écologie*, 1991.
10. R. Hippargi, P. Bolde, and A. Bolde, 'Ecological Importance of Footprint and Dust Bath Imprints of Great Indian Bustard (Ardeotis nigriceps)', *Ela Journal of Forestry and Wildlife*, vol. 5, 2016, pp. 199–205.
11. Dutta, Rahmani and Jhala, 'Running out of Time?'.
12. R. Manakadan, 'Ecology of the Great Indian Bustard Ardeotis nigriceps (Vigors) Habitats', MSc thesis, University of Bombay, 1986; A.R. Rahmani, A.R., *The Great Indian Bustard*, Final Report in the study of ecology of certain endangered species of wildlife and their habitats, Mumbai: Bombay Natural History Society, 1989.
13. A K-selected species, also called K-strategist species, are those whose

populations fluctuate at or near the carrying capacity (*K*) of the environment in which they reside. *K*-selected species are characterized by long gestation periods lasting several months, slow maturation (and thus extended parental care), and long life spans. See 'K.Selected Species', *Encyclopedia Brittanica*, see https://www.britannica.com/science/K-selected-species, accessed 25 Jnauary 2023.

14. A survivorship curve is a graphical representation of how many individuals in a species population are alive at different ages of life. A Type I survivorship curve suggests that individuals have a high probability of surviving through early and middle life but have a rapid decline in late life (for example, humans, elephants). This is depicted as a typical convex-shaped curve on the graph. A Type II survivorship curve represents a roughly constant mortality rate for the species throughout its entire life (for example, songbirds, squirrels). This is depicted through a diagonal downward straight-line pattern on the graph.
15. B.J. Tigar and P.E. Osborne, 'Invertebrate Diet of the Houbara Bustard Chlamydotis [undulata] Macqueenii in Abu Dhabi from Calibrated Faecal Analysis', *Ibis*, vol. 142, 2000, pp. 466–75; G.S. Bhardwaj et al., 'Role of Herbaceous Vegetation in Habitat Utilization by Critically Endangered Great Indian Bustard Ardeotis nigriceps (Vigors) in the Indian Thar Desert', *Indian Forester*, vol. 143, 2017, pp. 975–84; B. Bhushan and A.R. Rahmani, 'Food and Feeding Behaviour of the Great Britain Bustard Ardeotis nigriceps (Vigors)', *Journal of the Bombay Natural History Society*, vol. 89, 1992, pp. 27–40.
16. Blanford, *Fauna of British India, including Ceylon and Burma*.
17. Rahmani, *The Great Indian Bustard*; BirdLife International, *Threatened Birds of Asia: The BirdLife International Red Data Book*, Cambridge: Birdlife International, 2001.
18. A.R. Rahmani, 'The Godawan Saga: Great Indian Bustards in Decline', *Sanctuary*, vol. 21, 2001, pp. 24–8.
19. A.A. Khan et al, 'Status, threats and conservation of the Great Indian Bustard Ardeotis nigriceps (Vigors) in Pakistan', *Current Science*, 2008, pp. 1079–82; A.R. Rahmani, 'Need to Start Project Bustards', *Bombay Natural History Society, Mumbai*, vol. 20, 2006.
20. Khan, 'Status, Threats And Conservation of the Great Indian Bustard'.
21. Rahmani, *The Great Indian Bustard*.
22. Ibid.; Johnsgard, *Bustards, hemipodes, and sandgrouse*; Dutta, Rahmani and Jhala, 'Running Out of Time?'
23. Rahmani, *The Great Indian Bustard*; Dutta, Rahmani and Jhala, 'Running Out of Time?'

24. Rahmani, *The Great Indian Bustard*; Dutta, Rahmani and Jhala, 'Running Out of Time?'; A.R. Rahmani, *Status of the Great Indian Bustard in Rajasthan-I*, Bombay Natural History Society, 1986.
25. Rahmani, *The Great Indian Bustard*.
26. Dutta, Rahmani and Jhala, 'Running Out of Time?'; T.N. Gray et al., 'Habitat Preferences of a Globally Threatened Bustard Provide Support for Community-based Conservation in Cambodia', *Biological Conservation*, vol. 138, 2007, pp. 341–50; J. García et al., 'Spatial Analysis of Habitat Quality in a Fragmented Population of Little Bustard (Tetrax tetrax): Implications for Conservation', *Biological Conservation*, vol. 137, 2007, pp. 45–56.
27. R. Dharmakumarsinhji, *Study of the Great Indian Bustard*, Final report, Morges: WWF, 1971.
28. S. Ali, 'The Moghul Emperors of India as Naturalists and Sportsmen', *Journal of the Bombay Natural History Society*, vol. 31, 1927, pp. 833–61; A.O. Hume and C.H.T. Marshal, *The Game Birds of India, Burmah and Ceylon*, vol. I, 1878; repr., New Delhi: Bhavna Books and Prints.
29. P.D. Goriup and H. Vardhan, *International Symposium on Bustards*, Jaipur: Tourism and Wildlife Society of India, 1980.
30. S. Dutta et al., *Guidelines for Preparation of State Action Plan for Resident Bustards*, New Delhi: Ministry of Environment and Forests, Government of India, 2013.

7

Conserving the Last Javan Rhino in the World

Inov Sectionov

> You can't save the rhinos and you can't preserve a culture. I'm very pessimistic. Once it's gone, it's over.
>
> —Paul Theroux

INDONESIA IS HOME to two rhinoceros species: the Javan Rhino (*Rhinoceros sondaicus*) and the Sumatran Rhino (*Dicerorhinus sumatrensis*). The Javan rhino is very similar in appearance to the closely-related Greater One-horned Rhinoceros, but has a much smaller head and less apparent skin folds. Being large megavertebra, and because of their dominant role in the ecosystems where they occur, rhinos are recognized as 'key species' in biodiversity conservation. Their conservation will help to protect habitats ranging from lowlands to montane areas which benefit many other wildlife species as well. The term 'umbrella' species is often used to denote this function of providing protection for many other species in the ecosystem. For the same reason, they are also iconic and are widely used as 'flagships' of protected areas in which they occur and to attract attention for conservation from the general public.

PLATE 7.1: Javan rhino in Ujung Kulon National Park
Image courtesy: Camera trap Ujung Kulon National Park

The Javan rhinoceros (henceforth JR) is the rarest species of rhino. The last remaining wild population is found only in Ujung Kulon National Park (UKNP), Banten, Indonesia. The last census of JR by UKNP in 2019 mentioned that seventy-two JRs (*Rhinoceros sondaicus*) remain on the planet,[1] and as one of the rarest mega-herbivores on earth, they have maintained a single population in their last stronghold in Indonesia.[2] In addition, the JR is the most threatened of the five rhino species in the world and has been listed as Critically Endangered on the IUCN Red List.

HISTORY OF THE JAVAN RHINO IN UJUNG KULON

Being a descendant of ancestors that once inhabited the Earth some 20 million years ago, the JR once roamed around most of South–South East Asia from the Brahmaputra River valley in Assam (India) down to Sumatra and Java, and was still abundance in Sumatra and Java until the eighteenth century. In the island of Java, until the year 1900, the JR was still common as reported by Franck (1934),[3] who with the help of tracks estimated their number to be around 6 in south Garut, west Java. In 1934 a rhino was found living in solitary

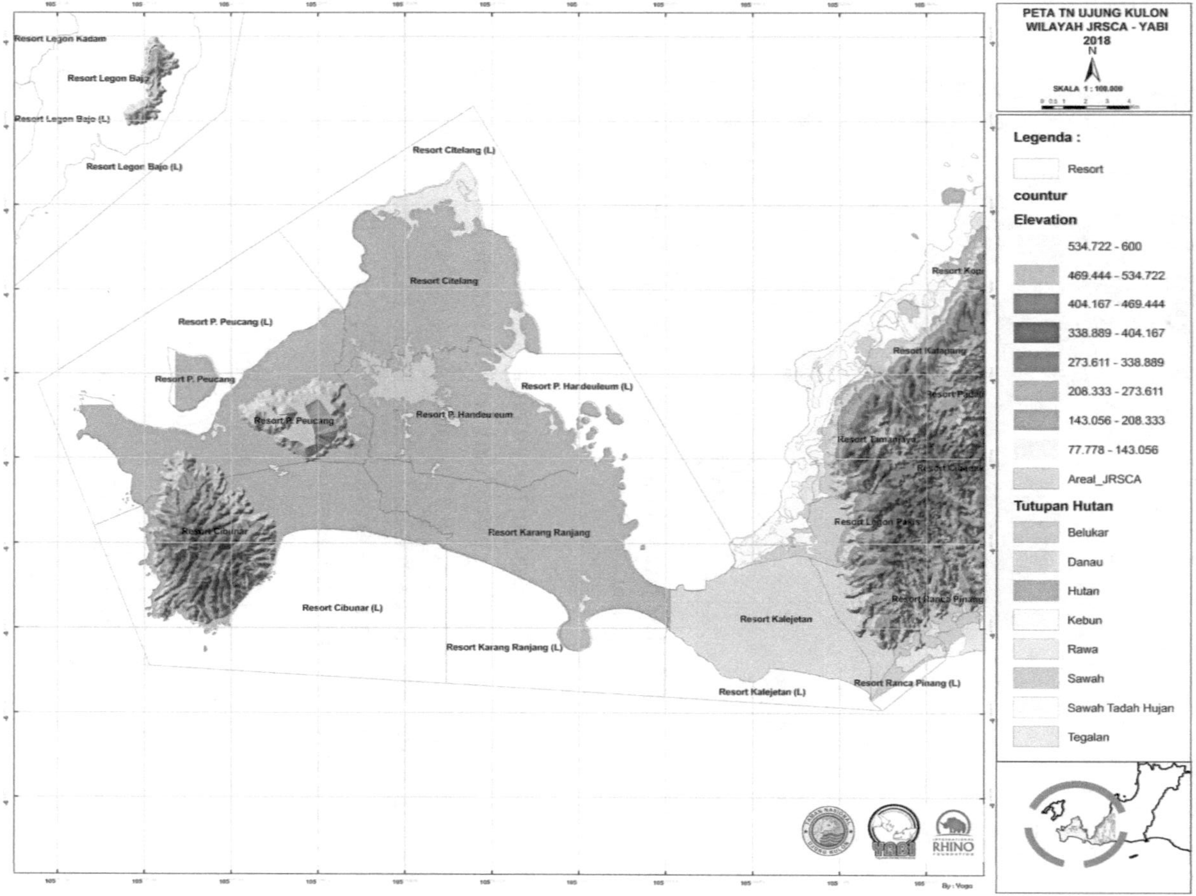

MAP 7.1: Ujung Kulon National Park in Indonesia. *Image courtesy*: YABI/IRF/UKNP

PLATE 7.2: The last Javan rhino killed outside UKNP in 1934
Image courtesy: Rhino Research Center

for more than five years in Karangnunggal Tasikmalaya; the animal was eventually shot dead to fill the Bogor Museum collection. According to F.J Appelman (1934),[4] this rhino was the last one that survived outside the Banten area. Since 1935 the JR can only be found in UKNP, aside from several individuals found in Mt Honje.[5]

THE DISAPPEARANCE

There was no clear explanation on the disappearance of this species from the rest of the Ujung Kulon area. However, it is clear that anthropogenic activities played as important factor for wildlife habitat decrease. Another important reason could be the fact that the JR is a browser-dependent species that needs proximity to human activities (which creates secondary vegetation as browser plants after slash and burn agriculture activities). This made the species vulnerable to human hunts. The need of lowlands for water to wallow

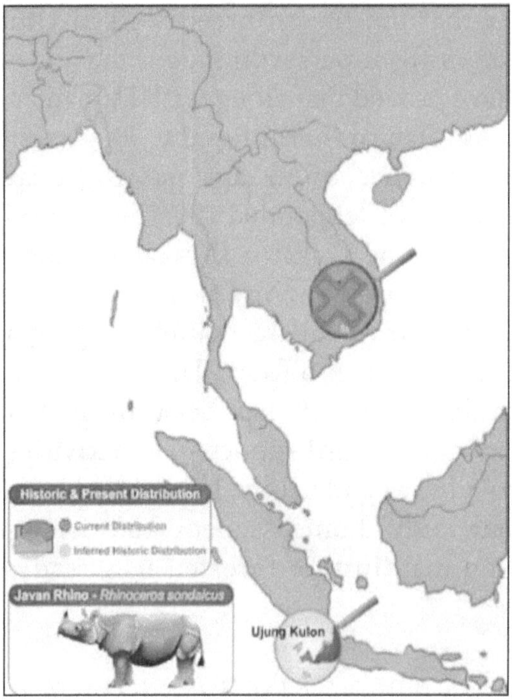

MAP 7.2: Historical distribution of the Javan Rhino
Image courtesy: International Rhino Foundation

and bath has been opined as yet another reason for the species coming in conflict with humans who also need the same terrain for agriculture and settlements. The growing human population thus pushed the rhino population to the margins and cornered them into the most western part of Java Island—Ujung Kulon.[6]

Rapid human population growth and development in Java, together with the opening of settlement areas and land conversion, further pushed the rhino aside. In 1909, the JR was designated a protected animal. In 1955, the number of JRs increased to thiry-five individuals, but the number once again declined in 1960 due to hunting. In 1967, twenty-one individuals remained inside Ujung Kulon. Since 1970, the rhino population steadily increased to forty-seven individuals and it reached fifty-one individuals by 1981. This growth demonstrates that the JR can breed well in Ujung Kulon if given protection from persecution.[7]

Studies have proved that along with human encroachment another serious factor harming the JR's existence is the increasing dominance of Arenga palms (*Arenga obtusifolia*). Arenga palms have expanded in the region since the 1883 explosion of the Krakatoa (Krakatau) volcano and now dominate the rainforest canopy in many areas of the park, substantially reducing available rhino foraging and limiting the growth of native rhino food plants.[8]

Thus, gradually the JR has been pushed to the brink of extinction due to anthropocentric activities (primarily hunting, agriculture) and habitat loss, sadly leaving the Ujung Kulon as their 'only' home. Genetic and demographic risks of in-breeding may further threaten its already debilitating population.

THE JAVAN RHINO'S FUNCTION IN THE ECOSYSTEM

Nature has, all of sudden, threatened the life of humankind on this earth. Or has this long-term process been induced by *Homo Sapiens?* Storms, high tidal waves, the instability of the seasons, etc., indicate some alarming recent changes. According to the latest data, the Earth's surface temperature has increased up to 3.5° celsius which has caused the polar caps to melt. Without reversal, it is estimated that in the next fifteen years, thousands of small islands will be forced underwater by the rising sea levels.

Drastic climate change has caused major habitat changes (mostly in the forests) which have a direct impact on the survival of its fauna. Changes in the duration of the dry or rainy seasons disrupt the behaviour of wild animals, causing stress as a result of changes in their natural habitat.[9] The animals cannot overcome and adapt to these changes fast enough, and neither can they deal with environmental change swiftly. Eventually, only the law of natural selection sustains and whoever can best adapt to change survives.

The JR is a big animal that evolved from its ancestors, the dinosaurs. However, in the Anthropocene, it has been critically affected by global warming. The changes in temperature and climate have altered its food plant's cycle and habitat. During the long dry season, the JR, which usually wallows in mud holes, is now languishing in search of mud holes due to paucity of water. Since everything in Nature is a part of chain, the wallowing process of the JR too plays a role in this ecosystem; the process is very effective in seed dispersal. Its ability to walk around many kilometers has been useful in spreading forest plants via the seeds it excretes in its feces. Many of these seed eventually germinate and grow into big strong trees with the roots grounded in the earth ready to serve as windbreaks, and to prevent floods and landslides.

CONSERVATION EFFORTS

In 1998, three teams of the Rhino Monitoring and Protection Unit were established as the frontline for ensuring the survival of the JR. These units have the task of protecting and conserving the rhino in its habitat and also conducting patrols (prevent area violation and poaching, find and destroy equipment threatening rhinos, confiscate and secure evident), law enforcement (conducting initial activity for law enforcement/investigation), survey and monitoring of rhino and other 'important' big mammals (such as banteng, deer, etc.), and intelligence missions (obtaining important information for law enforcement). The protection units were later merged with the Rhino Protection Units (RPUs).

The RPUs are a unique, highly professional collaboration, which bring government officers and local community members together to work towards a common goal. RPU operations are sustained with the support of the International Rhino Foundation (IRF) in collaboration with Ministry of Forestry Republic of Indonesia, while the RPUs day-to-day activities are managed under Yayasan Badak Indonesia

PLATE 7.3: Rhino Protection Unit or RPU
Image courtesy: International Rhino Foundation

(YABI). The RPUs provide security and management of the national parks in which they work by implementing regular patrolling.

The Action Plan for the Conservation of Rhinos in Indonesia (AsRSG) held a meeting in March 2009 to set immediate goals and actions in realizing the habitat quality in Ujung Kulon. It was estimated that Kulon could harbour up to a hundred JR, but suggestions were made to bring down the number as low as seventy. One of the strategies in the 2007 action plan was to create a second population of JRs to not keep all the eggs in one basket as such, and to increase habitats available to the rhino.[10] Also, to help mitigate the risk of extinction of the JR in UKNP, in 2010, authorities launched the 5,100 ha. Javan Rhino Study and Conservation Area (JRSCA) including installing an 8 km-long rhino-roof fence at the base of the eastern Gunung Honje Range to protect the habitat area, exclude domestic stock and protect the rhino.[11]

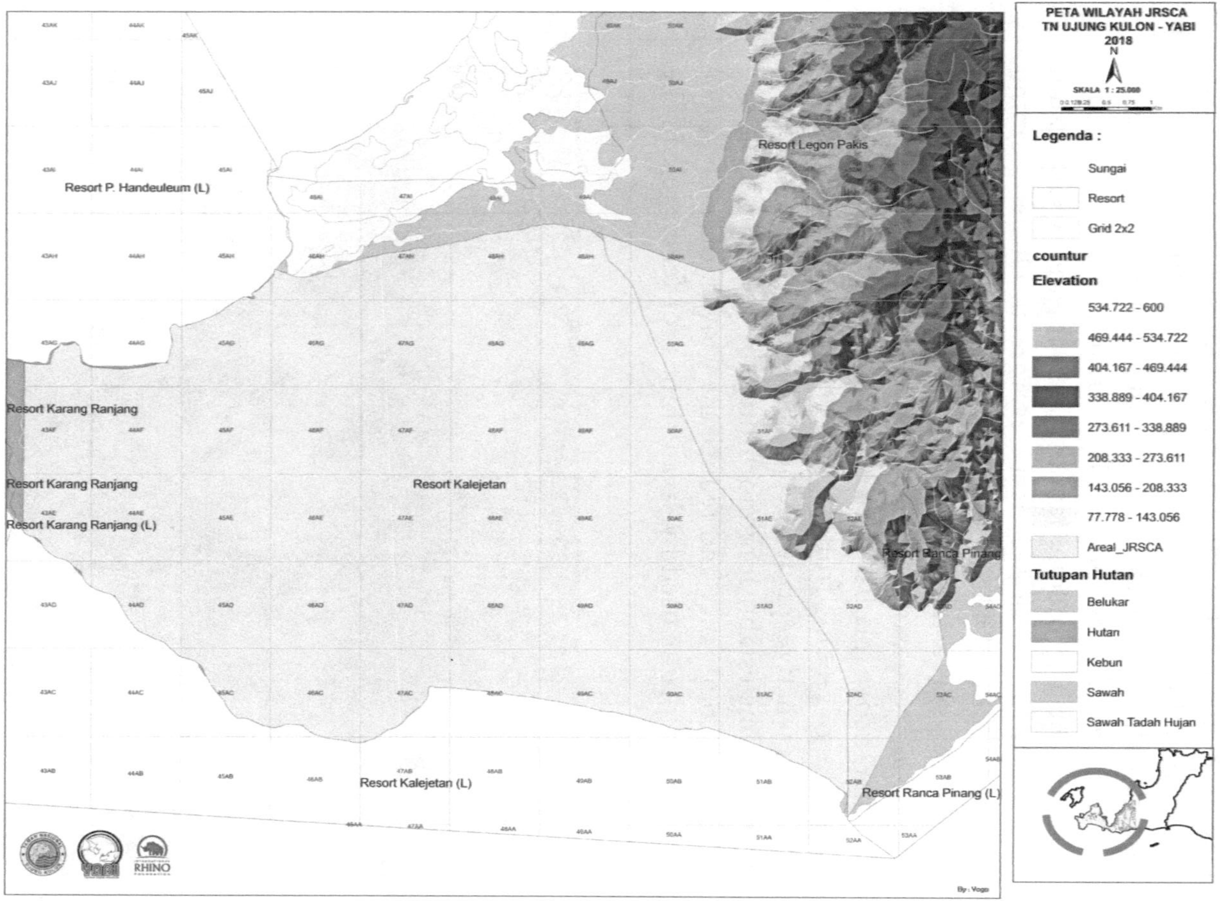

MAP 7.3: Map of Javan Rhino Study Conservation Area (JRSCA). *Image courtesy*: YABI/IRF/UKNP

MAJOR THREATS

One of the main threats to JRs is the limited suitable habitat, which is mitigated by the increasing dominance of the Arenga palm (*Arenga obtusifolia*) in the UKNP. A member of the Arecaceae family and widely distributed across Thailand, Malaysia and the Indonesian archipelago, the Arenga palm (locally known as *langkap*) is an evergreen, cluster-forming species that can grow to heights of 16 m.[12] Arenga palms spread by subterranean root growth, also termed root extension,[13] constraining and outcompeting the growth of other native plants.[14] The Arenga palms produce large quantities of seeding fruits, which enhances their dominance over other native plant species. Able to disperse rapidly, Arenga palms establish monocultures by reducing light due to canopy dominance and reduce the growth of understorey native food plants used by Javan rhinos for feeding.[15] In the

PLATE 7.4: Arenga palm in UKNP
Image courtesy: Inov Sectionov

PLATE 7.5: Javan rhino individual utilizes JRSCA
Image courtesy: YABI/IRF/UKNP

other words, where Arenga palms dominate, nothing else grows. Based on the analysis of satellite imagery from a 5,100 ha. area of the JRSCA, 1,730 ha. consist of Arenga palm (29.40 per cent).

The control of Arenga palm by the JRSCA is supported by the local people. Additionally, this has helped in instilling and augmenting an understanding of habitat management and the need to save the JR species that is nearing extinction. The JRSCA has helped to increase the JR population from only two rhinos to four to six individuals by the end of 2014, representing the first step toward increasing the habitat's carrying capacity for rhinos and resident population.

Threats also surface as the demand of rhino horns remains high and locals continue to collect firewood, herbs and other forest essentials, while rhino poaching remains a latent problem.

SECOND HABITAT FOR JAVAN RHINOS

From the beginning of the Indonesian Rhino Conservation Strategy (IRCS) in 1994 until today, the search for a second habitat for the JR has been repeatedly endorsed. Although the proposal and the transfer plans were initiated long ago, so far there remain no signs of implementation. Small and isolated populations are very vulnerable to accidental loss or poaching, to localized catastrophes like floods, fires, cyclones, tsunami and they can also suffer declining vigour or vitality through steady loss of genetic diversity. Intense discussions about how we can rescue the JR from imminent extinction threat continue.[16]

Transfer or translocation is not something new in wildlife conservation, particularly for rhinos. Rhino translocation has proven to be an essential tool for conservation of both the Black Rhino (*Diceros bicornis*) and White Rhino (*Cerathoterium simum*) in Africa. The Black Rhino population suffered severe poaching pressures in the 1970s and 1980s with significant decline in the global population. Today, the Black Rhino population is recovering thanks to translocation measures across the African continent. Moreover, translocation efforts have also begun to help secure preservation for the Greater One Horn Rhino or GOHR (*Rhinoceros unicornis*) in India and Nepal. Such conservation efforts are producing good results with high population growth rates of translocated GOHRs.

Learning from the African and Asian experience concerning translocation of rhinos, we feel these methodologies could be applied to the Javan Rhino and/or the Sumatran Rhino. Yet, we still must consider several factors important to any translocation activity, such as capturing techniques, transport techniques, habitat assessments, as well as critical planning and preparations to ensure success. Appropriate education and awareness programs aimed at various stakeholders will be essential. Translocation is one alternative to help rescue the JR from the threshold of extinction.

GLIMMERS OF HOPE FOR THE JAVAN RHINO

The critical condition of the JR defines the need of a well-orchestrated on-ground protection of its remaining population and habitat. The RPU, consisting of forest rangers and local communities, has taken long strides in this direction.

However, development of an area needs to go through public consultation in length and deter any hindrance to its aim. To answer such myriad problems, research activities by national and international researchers, universities and the national park regarding life, habitat, population and conservation of JR have been intensely carried out in sync with increasing awareness of the community around the JR habitat. Long strides have been taken and there are still miles left to go.

In addition, it has been demonstrated that careful habitat management by controlling the Arenga palm, developing a grid system, and providing protection by the JRSCA has helped the JR reside in the area. This has also helped in demarcating a safer wilderness for the JR and ensuring

PLATE 7.6: Fencing at the Javan Rhino Study Conservation Area (JRSCA)
Image courtesy: Inov Sectionov

protecting from human and cattle encroachments. Efforts are still in progress to search for a secure second habitat for the JR, in part, because a difference of opinion still exists among stakeholders about the most appropriate location especially from the point of view of security.

We hope for a future in which researchers, stakeholders and authorities are united and pledge towards creating a second safe haven for the majestic JR. This species needs to be saved not just for the biodiversity of Indonesia, but for Nature's invisible chain that sustains all of us. Aren't living animals more valuable than their stories from the past?

NOTES

1. B. Gokkon, 'Newly Spotted Calves Boost Javan Rhino Population to 72', *Mongabay* (website), 2019, see https://news.mongabay.com/2019/12/javan-rhino-calves-population-72-ujung-kulon-indonesia/, accessed 20 December 2022.
2. R.H. Emslie et al., 'African and Asian Rhinoceroses: Status, Conservation and Trade', Report from the IUCN Species Survival Commission (IUCN SSC) African and Asian Rhino Specialist Groups and Traffic to the CITES Secretariat pursuant to Resolution Conf.9.14 (Rev.CoP17), Columbo, June 2019, 2019, pp. 1–38.
3. P.F. Franck, 'Verslag over de jaren 1933–34', *Nederl.-Ind. Vereen. tot Natuurbescherming*, 1934, pp. 40–55.
4. F.J. Appelman, 'Rhinoceros sondaicus in West-Java', *Tropische Natuur*, vol. 23, 1934, pp. 73–6.
5. Inov Sectionov, 'Translocation rhino in Indonesia', Presentation to the 2007 Translocation Rhino Workshop in Kenya for Producing Guidelines for the In-situ Re-introduction and Translocation of African and Asian Rhinoceros, 2013.
6. Patrick W.F.M. Hommel, *Landscape-ecology of Ujung Kulon (West Java, Indonesia)*, n.p.: Patrick W.F.M. Hommel, 1987.
7. Patrick W.F.M. Hommel, 'A phytosociological study of a forest area in the humid tropics (Ujung Kulon, West Java, Indonesia)', *Vegetatio*, vol. 89, 1990, pp. 39–54.
8. Inov Sectionov, 'Control of Invasive Arenga Palm (*Arenga obtusifolia*) in Habitat Suitable for Javan Rhino (*Rhinoceros sondaicus*) in Ujung Kulon National Park', Presentation to the 2013 International

Elephant and Rhino Conservation and Research Symposium in Pittsburgh the USA, 2013.

9. Inov Sectionov, 'Rhinos and Global Warming', *Rhino Resources Center* (website), 2018, see http://www.rhinoresourcecenter.com/pdf_files/120/1202896893.pdf, accessed 20 December 2022.
10. 'Strategy and Action Plan for the Conservation of Rhinos in Indonesia', Directorate General of Forest Protection and Nature Conservation; Ministry of Forestry of the Republic of Indonesia, 2007.
11. S. Ellis, 'Enhancing the Survival of the Javan Rhino', *Endangered Species Bulletin*, Spring 2010, pp. 40–1.
12. '*Arenga obtusifolia* Palm', Global Invasive Species Database (website), see http://www.issg.org/database/species/ecology.asp?si=1834&fr=1&sts&sts=&lang=EN, accessed 17 February 2017.
13. A.R.S. Hariyadi et al., 'Optimizing the Habitat of the Javan Rhinoceros (*Rhinoceros sondaicus*) in Ujung Kulon National Park by Reducing the Invasive Palm *Arenga obtusifolia*', *Pachyderm*, vol. 52, 2012, pp. 49–54.
14. Supriatin, 'Studi kemungkinan adanya pengaruh alelopati langkap (*Arenga obtusifolia* Blumme ex. Mart) terhadap pertumbuhan semai tumbuhan pakan badak Jawa (*Rhinoceros sondaicus* Desmarest, 1822) di Taman Nasional Ujung Kulon', PhD dissertation, Institut Pertanian Bogor, Bogor, Indonesia, 2000.
15. M. Haryono et al., 'Monitoring of the Javan Rhino Population in Ujung Kulon National Park, Java', *Pachyderm*, vol. 56, 2015, pp. 1–5.
16. Muh. Haryono et al., eds., *Population and Habitat Viability Assessment for the Javan Rhino*, Apple Valley, MN: IUCN/SSC Conservation Breeding Specialist Group, 2016.

8

Orangutan
Person of the Forest

Ulhas Gondhali and S.K. Gupta

> Surely, we do not want to live in a world without the great apes, our closest living relatives in the animal kingdom?
>
> —Jane Goodall

Tracing the etymology of 'Orangutan' to its Malay origins, the word can be broken into two words—'Orang', which means 'person', and 'Hutan', which means forest. Thus, it translates to 'Person of the Forest'. Although the name has a Malay origin, it has never been used locally to denote this ape. The indigenous name for the orangutan is '*mawas*' in Sumatra or '*maias*' or '*mias*' in Borneo. It appears that Jacobus de Bondt is the first to have used the name 'orang-outan'.[1] Another possibility is that the word came from an aboriginal belief that orangutans were nothing but humans hiding in the forest to avoid working. Its intelligent behaviour and stark similarities with human culture may have led to this belief. However, today we have considerably improved our understanding of this fascinating animal.

Orangutans are arguably the most difficult animals to sight in their natural habitats. One might not have seen an orangutan in a zoo or on any wildlife show on TV, but it must have not gone unnoticed as a fictional character, King Louie, in Walt Disney's 1967 animated musical film, *The Jungle Book*. The character was portrayed as a Giant Orangutan who was a leader of other jungle primates.

Orangutans are the only great ape found in Asia; today, there are three distinct species reported—Sumatran Orangutan (*Pongo abelii*), Bornean Orangutan (*Pongo pygmaeus*), and Tapanuli Orangutan (*Pongo tapanuliensis*).[2] However, initially, it was thought that only one species existed, and the Sumatran Orangutan and Bornean Orangutan were considered subspecies of *Pongo pygmaeus*. In early 2000, after a taxonomic review, it was accepted that the Sumatran Orangutans are different from their Bornean relatives. Sumatran Orangutans are endemic to Sumatra, Indonesia, and Bornean Orangutans are endemic to Borneo, Indonesia, and Malaysia. In 2017, a third species was added, the Tapanuli Orangutan. This distinction was made based on its morphology (cranium and teeth) and genomic structure. Historically, orangutans were found all over South-East Asia, including Sumatra, Java, and China. These animals have been facing extensive difficulties from various anthropogenic activities such as deforestation and hunting. Currently, these species have been restricted to protected forest lands in Borneo and Sumatra. All three species are reported as 'Critically Endangered' by the IUCN.

TAXONOMY

Humans and orangutans share a common origin, the family Hominidae. The genus they belong to is *Pongo*. Some researchers argue that animals falling under *Pongo* should not be included in the Hominidae family because they do not share morphological and behavioural characteristics with humans. As mentioned before, orangutans belong to the genus *Pongo*, which is further divided into three species—Sumatran Orangutan, Bornean Orangutan, and Tapanuli Orangutan.[3] However, the Tapanuli Orangutan that is different in appearance and genetically isolated from its two other relatives.

The Bornean Orangutan is further divided into three subspecies:[4] *P. p. pygmaeus, P. p. wurmbii and* P. p. *morio.*

MORPHOLOGY

The difference between African apes and orangutans was realized by the Western world at the beginning of the nineteenth century.[5] Prior to this, researchers were mostly looking at the great apes to find connections between humans and other animals. Various studies were performed in the nineteenth and early twentieth centuries on orangutans to understand their morphology and anatomy. Scientists became less interested into studying orangutans after the mid-twentieth century because of popular opinion that chimpanzees were the closest relatives to humans. In the last decade, the discovery of hominids fossils and comparison of the taxonomic relationship among Hominoidea has once again generated interest among researchers of the orangutan.

The orangutan is the world's largest tree-dwelling animal. They have a large body that is covered by a thin layer of long, dark brown or orange hair. Due to this distinct colour, their movement on the tree tops is visible. However, the nature of their noticeable appearance has not been understood till now. They have very long and strong arms, even longer than their legs, which help them in moving between trees, as they spend much of their time atop trees. Their legs have an excellent grip which allows them to climb effortlessly. They have long fingers and operational thumbs like humans that can touch their other fingers, enabling them to carry out complex tasks with their hands. However, like other apes, they do not have tails.

Orangutans display sexual dimorphism—a distinct difference in appearance between males and females. Males, during their mature period, are two to three times heavier and considerably bigger than females. Researchers have also

PLATE 8.1: An adult male orangutan in the forests of Asia
Image courtesy: Jeremy Bishop (Unsplash.com)

recently confirmed bimaturis—two distinct types of sexually mature males, namely flanged and unflanged. The fully developed flanged males have been named so because they have a flange or cheek pads made of subcutaneous fibrous tissue. They have throat pouches that are used for making long calls. On the other hand, the unflanged males, also called 'arrested males', do not develop flanges. Their growth of secondary sexual characteristics gets arrested. However, the difference in appearance in the adult males has no effect on their sexual life.

The interspecies difference in orangutans is also evident in their morphology. Bornean Orangutans have rounder faces as compared to their Sumatran relatives, who have a longer face. There is also a colour difference in their hair; Bornean Orangutans vary in colour from orange/red to chocolate brown. Sumatran Orangutans have brighter orange hair.

Orangutan's longevity in the wild appears to be around thirty-five to forty years; in captivity, they could live for more than fifty years. Sexual maturity comes around the age of 7–10 years for females, though they typically do not give birth until 12–15 years of age. Males can be fertile as early as 7 years of age, but they do not show secondary sex characteristics until 10–15 years of age. In mammals, the O=orangutan has the longest birth interval. The interval can range from eight to ten years. The high age of breeding in females and long birth intervals are concerns for animal conservationists. Once a female is lost, it takes considerably long to recover a generation.

ECOLOGY

Orangutans are primarily tree-dwelling animals. Historically, they they were found across South-East Asia, but now they are confined to the tropical rainforests of Borneo and Sumatra. Considering its historical distribution range, they are adept at living in several different primary forest types, including swamps and lowland forests at sea level, to hilly forests at higher altitudes.[6]

Due to serious habitat loss, currently Sumatran Orangutans have been restricted to the northern tip of Sumatra in fragmented forests. The fate of many such species living in South-East Asia's rainforests is in grave danger because of habitat destruction, excessive hunting, forest degradation, and fragmentation.

SOCIAL BEHAVIOUR AND CULTURE

When we talk about 'culture', we usually associate it with humans. As we travel over different human settlements in the world, we see cultures proliferating. There can be different culture-specific ways of communicating, different food habits, etc., due to cultural/racial divergence. This concept of culture is not just limited to humans. Cultural models can also be seen seen in the animal kingdom, and it is not something that has been recently discovered. Researchers have been studying cultural practices among orangutans for a long time. Extended research has established that there are cultural variations between different population groups as well as culturally specific behavioural patterns.[7]

Richard Dawkins, in his book *The Selfish Genes* has defined basic cultural elements called 'memes'.[8] A meme is an innovation or idea that can be transmitted or communicated from one individual to another. These can also be copied by other individuals. Some memes are transmitted from one generation to another. Over time, as memes passed on through generations, they define cultural elements of a group of individuals.

Memes and culture have also been observed in orangutans. One good example of such memes is their umbrella-making skill. When it rains, an orangutan will quickly make an umbrella out of leaves and twigs. It's unlikely that this behaviour is genetic. A possible explanation for this is that they could have learnt this behaviour watching fellow orangutans making umbrellas. This suggests that at some point in the past, one ingenious orangutan invented an umbrella and the behaviour was passed on across generations. Today, this umbrella making is adopted by all orangutan populations.

PLATE 8.2: A mama orangutan holding her baby in Memphis Zoo, USA
Image courtesy: Jeremy Bishop (Unsplash.com)

However, since these populations are fragmented, each could portray minor differences in the umbrella-making process.

In a research study conducted in the forest of Gunug Palung by biological anthropologist Cheryl Knott and field biologist Robert Rodriguez, some interesting observations were made.[9] They followed a male orangutan called Codet. After several days, they started noticing some peculiar behaviours in Codet, which they had never observed in any other irangutan. One day, while eating leaves from a small tree, he ripped a branch off a tree and stuck it on his head. It looked like a hat! He walked around for some time and then stopped, took the branch off his head and took a bite. He then, once again, stuck the branch on his head and continued walking. He seemed to have figured out a way to take his food on the go and keep his hands free. This is something that Codet could have learnt from another orangutan or devised all by himself. This could be a 'meme' in the making, with chances of spreading to other natives.

WHY SHOULD WE CARE ABOUT ORANGUTANS?

The orangutan is an umbrella species. Conservation plans designed for it also entend protection to other species aound them. As orangutans need large forest areas to survive, by conserving their habitats many other plant and animal species living there are also protected. Thus, all conservation efforts for a species like this have a widespread or protective umbrella effect on numerous other co-occurring species. Generally, in wildlife conservation planning, one such species which is spread across a large territory is selected. Channelizing conservation efforts on one such species reduces the relative cost and, simultaneously, it can benefit several other co-occurring species. Male orangutans cover a large territory; they can travel a thousand metres per day in search of food. They need large spaces to prosper. As a result, protecting areas like the Betung Kerihun National Park, spread over 8,000 km^2 areas, benefits every species in that space.

The Orangutan is also a keystone species. A keystone species refers to a group of animals whose actions govern the well-being of a large majority of species in its given ecosystem. An orangutan's diet consists primarily of fruits. They forage most of their diet from over a hundred different species of fruit trees. As a result, they disperse seeds everywhere. Orangutans help to shape habitats and provide food for other fruit-eating species.

There are millions of species present in the world. Scientists are tirelessly working towards their classification, protection and conservation. However, the common masses are aware of only a few hundred of them. A flagship species is one which is iconic because of its unique appeal. There are many such flagship species globally, such as tigers, panda bears, polar bears, etc. Orangutans are another such species that is classified as a flagship species. The prime objective of flagship species is to take advantage of their unique appeal to raise awareness for animal conservation. Often, such species are the face of many animal conservation efforts. Orangutans are the face of conservation efforts in areas like Sumatra and Borneo. The areas where orangutans reside today are biological hotspots. These areas are covered with dense forests, which are home to thousands of wild species. Unfortunately, these forests are continuously being destroyed due to human activities, which has a clear negative effect on the local wildlife, including orangutans. Many organizations are working to minimize and eventually stop this destruction of nature. Being a flagship species, orangutans draw the attention of the rest of the world to this ongoing issue. Thus, orangutan conservation can also have a positive effect on many other local species.

CONSERVATION

To keep up with the ever-growing demands of the twenty-first century, many countries are over-exploiting their natural resources. The race towards development has caused many irreversible damages to the environment. Modern societies are built at the cost of nature. Contrary to that, the primitive societies—such as the Adivasis of India or Aboriginals of Australia—tend to follow practices that preserve nature. For example, the fishermen community around the southern coast of India follow traditional practices that are passed down from generation to generation. They are aware that over-exploitation and the ever-changing effects of climate change affect fish in the local waters. Thus, many of them have been moved to sustainable fishing methods.

Wildlife conservation has a dual purpose. The primary one is the conservation of species and their diversity; the other purpose is animal welfare. Animal welfare is targeted to reduce sufferings of animals in the human care.

POPULATION TRENDS

The population of all three species of orangutans is declining; the long-term sustainability of the entire genus is in question. The Sumatran Orangutans are facing immediate danger of extinction. Recent studies estimated a decline of 25 per cent in the population of Bornean Orangutans over ten years (2000–10) and approximately 50 per cent in the period 1999–2015.[10] Based on predicted land cover loss, 43,000 Bornean Orangutans could be lost by 2050. The decline of Sumatran Orangutans from its population of 13,900 in 2016 has not been systematically quantified. However, if we relate it to forest loss data, the decline is very steep. Between 1985 and 2007, about 60 per cent of the primary forest areas of Sumatran Orangutans were lost.[11] Future predictions indicate further loss of about 4,000 Sumatran Orangutans by 2030. Because the Tapanuli Orangutans were discovered very recently, the population trends are not known, but a quantitative population analysis study estimates that in 1985 there were approximately 1,489 Tapanuli Orangutans.[12] If the present threats to orangutans are not reduced and/or controlled, then 83 per cent of the population Tapanuli Orangutan may be lost.[13]

EVOLUTIONARY HISTORY

A study on orangutan DNA has provided evidence about the present species and their demographic distribution. Various evolutionary forces have shaped the history of Bornean and Sumatran Orangutan populations, including biotic and abiotic factors. Three prominent elements essential to understanding this evolution are: biogeography of the region over the last million years, human impact, and the social structure of orangutans.[14]

There have been many climatic and geographical changes in the region. Recurring glacial periods led to climatic changes and contracting rainforests isolated orangutan populations. The drop in sea level created bridges between islands, therefore creating opportunities for migration. Rivers and mountain ridges lead to dispersion as they are geographical hurdles for the species, dividing their natural habitat and leading to fragmentation of the population.[15]

The social structure of orangutans is also different from other primates, which could have contributed to variation at the DNA level.[16] They live in a semi-solitary social system characterized by a loosely organized group of solitary females with their young ones. They also exhibit sex-biased dispersal; males often leave their birthplaces, but females stay connected to theirs. Besides, as discussed earlier, females show a strong preference for dominant males with a flange. This sex-biased dispersal and reproductive choice has had an impact on genetics.

Orangutans have been affected by hunting and other anthropogenic activities such as farming since ancient times. There is a drastic increase in recent human activities (hunting, aforestation, pet trade), leading to a dramatic reduction in population size. Rampant destruction of the native forests of Sumatra and Borneo has resulted in the depletion of resources and habitat for orangutans. However, the precise impact of human-led activities has not been studied at the genetic level.[17]

THREATS

The major threats to orangutans are hunting for meat and other reasons, habitat destruction, and fragmentation of forests. The influence of these threats varies based on location. In areas where the human population is dense, habitat loss is higher with very few orangutans found. The adults are poached for their meat and the accompanying infants are sold in local and international markets.

MEN VS. ORANGUTAN: HUNTING

As discussed earlier with regard to the population trends of orangutans, one of the reasons for the steady decline in orangutan populations is hunting. Orangutans have been hunted for centuries. Although a recent major contributor to the eradication of the species is habitat destruction, hunting has a heavy yearly toll on the population. There is indirect evidence on hunting by indigenous people in accounts written by various authors until industrialization.[18] Fossils found in the caves of Niah in Malaysia are proof that man has been hunting orangutans for 40,000 years. Hunting for food and other necessities was, and still is, a part of the culture in Indonesia and Malaysia. Hunting by locals is one of the most complex issues faced by conservationists as they do not want to offend locals or disrupt local and native practices.

Hunting has played a vital role in the evolution of the modern human. Hunting of closely related species to humans or primates could suggest predatory behaviour on the part of humans.[19] In other words, this behaviour can be closely related to headhunting or cannibalism. Poaching is synonymous with the illegal killing of animals. The purpose of hunting can be for the consumption of bushmeat, trophy hunting, or retaliation for crop foraging. Bushmeat is any wild animal meat consumed by humans. Bornean Orangutans are killed in large numbers for their meat or when they are threats to human safety or crops. Sumatran and Tapanuli

Orangutans are also on a similar scale of risk and are hunted for meat or killed as a result of human–animal conflicts. However, a high level of hunting is obseved in Borneo; this is comparatively lower in regions inhabited by the Sumatran or Tapanuli Orangutan. In the current age of development, new roads connecting remote parts of the forests enable easy access and bushmeat transport for illegal trade.

Owning or trading Ooangutans in Indonesia and Malaysia has been illegal for decades; however, this has not stopped the animal trade. Illicit trade is one of the most severe threats to the conservation of orangutans. Orangutans are listed in Appendix I of the Convention on International Trade in Endangered Species (CITES). Indonesia and Malaysia are both signatories of this body and they must adhere to this prohibition. There are also national laws in Indonesia and Malaysia for the protection of orangutans. The laws were initially introduced by Europeans during the colonial period and were later amended after Independence. Both the countries are also members of the ASEAN Wildlife Enforcement Network (ASEAN-WEN). Indonesia is bound by the regulations of the Kinshasha Declaration to improve 'the protection of individual great apes and their habitats everywhere by demonstrably improving where necessary the quality and enforcement of relevant laws, as well as the capacity of law enforcement agencies'.[20] Despite many such efforts on monitoring and stopping illegal trade, the picture today is still dismal. Today, the illegal wildlife trade market is the second-most lucrative market (the first being the international drug tade), with a valuation of around 10 billion dollars annually.[21]

Primates are popular animals to sell at the international level. They are traded for many reasons, including:

1. To be used as pets
2. To use as experiments for medical research
3. For zoos and the entertainment industry[22]

Human encroachment into forests has been a major contributor to human–wildlife conflicts. As their habitats are destroyed and fragmented, orangutans are forced to leave their homes and live near human settlements. Usually, loggers or field workers are the ones who encounter wild orangutans leaving the forests. These generally include mother orangutans and their children. In such encounters, humans often attack the animals, which results in the death of adults for meat and the capture of the younger ones to be traded as pets. The illegal sale of young orangutans is a global issue. Monetary values of orangutans differ from place to place. For example, in Kalimantan, a baby orangutan can fetch anything from $27 to $112. In the markets of Java, orangutans cost more than $385. Sometimes orangutans are also exchanged for other commodities like electronic appliances or wristwatches.[23] When sold internationally, the younger ones are often kept in terrible conditions to evade customs officers. Most of the young orangutans don't survive the cruel journey to the wildlife markets or private pet owners' homes in a distant part of the world. After they reach their destination in poor conditions, diseases and the mental trauma of losing their mother further degrade their condition. Because of such mental and physical trauma, most young orangutans don't survive their first year. It is estimated that the ratio of survival of orangutans in such conditions is 1:8,[24] which means, on average, eight orangutans die for the survival of one. If they survive through initial difficulties for around two to three years, they will be traded in the pet industry. However, after this initial period of about four years, they grow to become as strong as an adult human, and by the time they hit maturity around the age of 8 to 10, they become many times stronger than an adult human. At this age, due to their size and physical power, they become difficult to handle. As a result, most owners either cage them permanently, kill them or abandon them. Because of this, there are many rehabilitation programmes for orangutans in Borneo and Sumatra.

PLATE 8.3: A caged orangutan looks out with sadness in its eyes
Image courtesy: Jeremy Bishop (Unsplash.com)

Orangutans are also captured to be kept in zoos and wildlife parks. Often the records of animals kept in the zoo or similar places are uncertain; the place of origin of animals is often unknown. Few are likely to be captive-born. The conditions in zoos are usually terrible with high mortality rates. In recent years there has been growing demands for orangutans in the entertainment industry—they are used in highly publicized kickboxing matches. In 2004, around 114 orangutans were rescued from the Safari World Theme Park in Bangkok as they were forced to dress up and engage in kickboxing. In 2006, 54 Orangutans were moved to Indonesia and the rest are believed to have died.[25]

ORANGUTANS AND THE PALM OIL CRISIS

Today, palm oil is one of the most commonly used vegetable oils in the world. It is found in more than half the processed products produced worldwide, from cookies to cosmetics. This massive consumption of palm oil has created an enormous demand in the international market. Palm trees grow in tropical areas and need a lot of water. They grown over South America, West Africa and South-East Asia. Indonesia is one of the biggest exporters of palm oil and the country earns 1/10th of its total export earnings from palm oil.[26] To meet growing demands, countries like Indonesia are pushing for higher rates of palm oil production. The problem with the palm oil industry in Indonesia is that a huge portion of forest land is being cleared out for palm plantations. This industry is leading to forest destruction in South-East Asia, and countries like Indonesia and Malaysia are majorly affected. In Sumatra, millions of hectares of forest are being opened to the palm oil industry. The conditions in Borneo are not very different. Clearing out land that is home to millions of animals has a devastating effect on local biodiversity. It is also a significant threat for orangutans as their natural habitat is being destroyed for palm plantations. As a result, orangutans are being forced to roam around in search of livable habitats. They often enter palm fields and end up damaging crops. When they repeatedly encounter humans, it results in human–wildlife conflict, which has severe effects on orangutan conservation as locals then begin treating orangutans as pests.

WAYS TO SAVE ORANGUTANS FROM GOING EXTINCT

1. Shop cautiously: In the age of globalization, many products travel halfway around the world to reach consumers. These products have a huge carbon footprint and are often made unsustainably. While purchasing products, from groceries to stationery, make sure to buy locally made

PLATE 8.4: A mama orangutan with her baby in a forest of Asia
Image courtesy: Jeremy Bishop (Unsplash.com)

or locally sourced products. A majority of products now contain palm oil, which is often disguised as vegetable oil. The production of palm oil comes at the cost of the environment as huge chunks of rainforests are cut down for palm plantations. Avoid products containing palm oil. Raise your concerns about the production companies' overuse and procurement of unsustainably sourced palm oil. For products made of paper or wood, check for the mark of the 'Forest Stewardship Council'. These products come from sustainable forestry.

2. Make small lifestyle changes: Avoid using single-use items. Use environmentally friendly and recyclable items. Wherever possible, use recycled products. Such small changes can have a positive impact on the environment.
3. Raise awareness: People are not aware of their everyday consumer choices and their impact on the environment. Educating people around us on sustainability and animal conservation will help them understand this global issue.

Often, people have misconceptions about sustainable choices. Lead them by example.

Lets us, the human race, not be our closest relatives' greatest enemy. And let us not forget the fact that, it can lead other species to extinction, humankind is sure to follow suit.

NOTES

1. H.D. Rijksen, *A Field Study on Sumatran Orang Utans (Pongo pygmaeus abelii, Lesson 1827): Ecology, Behaviour and Conservation*, Wageningen: Veenman, 1978.
2. I. Singleton et al., *Orangutan Population and Habitat Viability Assessment: Final Report*, Apple Valley, MN: IUCN/SSC Conservation Breeding Specialist Group, 2004.
3. Ibid.
4. N. Bemment, *Orang utan EEP Best Practice Guidelines*, vol. 1, 1st edn., ed. Neil Bemment, Barcelona: Great Ape Taxon Advisory Group, 2019, https://www.eaza.net/assets/Uploads/CCC/OU-EEP-Best-Practice-Guidelines-final.pdf, accessed 20 December 2022.
5. Rijksen, *Field Study on Sumatran Orang Utans*.
6. Singleton, *Orangutan Population and Habitat Viability Assessment*.
7. Cannon, J., 'Orangutan Culture in Focus in "Person of the Forest": Q&A with Researchers Cheryl Knott and Robert Rodriguez Suro', *Mongabay* (website), 23 February 2018, see https://news.mongabay.com/2018/02/orangutan-culture-in-focus-in-person-of-the-forest-qa-with-researchers-cheryl-knott-and-robert-rodriguez-suro/, accessed 20 December 2022.
8. Ridley, M., 'In Retrospect: The Selfish Gene', *Nature*, vol. 529, no. 7587, 2016, pp. 462–3.
9. Cannon, 'Orangutan Culture in Focus in "Person of the Forest"'.
10. D. Greer et al, 'Great Apes Status Report (August 2018)', Report to the CITES Standing Committee on the Status of Great Apes, United Nations Environment Programme Great Apes Survival Partnership, Nairobi, and International Union for Conservation of Nature, Gland, 2018.
11. Ibid.
12. Ibid.
13. Ibid.
14. X. Ma et al., 'Population Genomic Analysis Reveals a Rich Speciation

and Demographic History of Orang-utans (Pongo pygmaeus and Pongo abelii)', *PLoS One*, vol. 8, no. 10, 2013.

15. Ibid.
16. B. Goossens et al., 'Patterns of Genetic Diversity and Migration in Increasingly Fragmented and Declining Orang-utan (Pongo pygmaeus) Populations from Sabah, Malaysia', *Molecular Ecology*, vol. 14, no. 2, 2005, pp. 441–56.
17. Ibid.
18. Rijksen, *A Field Study on Sumatran Orang Utans*.
19. Ibid.
20. *CITES/GRASP Orang-utan Technical Mission Indonesia*, Geneva: CITES, 2006, see https://cites.org/sites/default/files/common/prog/ape/ID_mission06.pdf, accessed 20 December 2022.
21. G. Vince, 'Organized Gangs Target Wildlife Trade', *New Scientist*, 2002.
22. V. Nijman, *An Assessment of Trade in Gibbons and Orang-utans in Sumatra, Indonesia*, Selangor: TRAFFIC Southeast Asia, 2009.
23. Ibid.
24. Ibid.
25. Vince, 'Organized Gangs Target Wildlife Trade'.
26. K.R. Szulczyk, 'The Economics of the Malaysian Palm Oil Industry and its Biodiesel Potential', *SSRN Electronic Version*, 2013, pp. 1–18.

9

Amur Leopard

The Far Eastern Solitary Prowler

Pronami Bhattacharyya

> In the time it takes you to read this page, one of our planet's unique species will become extinct. By this time tomorrow, a further 150–200 will have disappeared forever. And by this time next year, over 50,000 more.
>
> —WWF Priority Species

I fostered a profound love for the big cat family when I was 4 years old. My father was posted in Nahorhabi Tea Estate, Assam (India). On a summer day in 1990, I learnt of the lynching of a *bagh* by some locals, who dumped its body at the nearest police station. A photo of the beautiful animal lying lifeless circulated widely. I never understood why it was killed. Even as a child, I asked my father, 'Could it not be captured by any means and returned to its family?' He had no answer. That was a time when I was not aware of the taxonomy of the animal or even that it was a leopard. I heard people talking about a *bagh* all the while; *bagh* is an Indian term for tiger. So, a *bagh* it was, only with spots and not stripes! Lack of awareness is the cog that has taken the largest toll on nature.

Of the big cat family, one of the rarest sights in the wild is of the Amur Leopard (*Panthera pardus orientalis*). A subspecies, it is also known as the Far Eastern Leopard, Manchurian Leopard, Korean Leopard, and Siberian Leopard. It was almost led to the brink of extinction due to mindless anthropocentric activities. It is one thing to pluck a beautiful flower, but it is altogether an abomination to wipe out a species. The population of the Amur Leopard reached a shocking low of thirty individuals in the wild in 2007.[1] However, dedicated conservation work has helped nearly double the numbers in the last few years. A 2022 report shows the number to be 70 individuals in the wild.[2] But it is still critically endangered and we have miles to go before we can be unburdened of this guilt.

THE RED FACTS

Among the big cats, the Amur Leopard has a special place, having been catalogued in the IUCN Red List as Critically Endangered[3] and in Appendix I on the CITES (the Convention on International Trade in Endangered Species).[4] The residual population in the wild, estimated at fifty-seven, is confined to a small area in the Russian Province of Primorsky Krai, between Vladivostok and the Chinese border.[5] Since the 1970s, its range has shrunk drastically, and in the late 1990s, the Amur Leopard vanished from most parts of north-east China.[6]

A research shows that there are only fourteen to twenty adults and five to six cubs in the south-western Primorye region of Russia; some may still occur in North Korea, although their status there is indeterminate; Less than ten leopards likely existed in Jilin and Heilongjiang provinces of north-east China towards the later part of 1990s, during which time they almost vanished. However, the population seems to have improved a little in recent years (42 leopards were photographed, including 40 adults, 21 males, 17 females, and 2 of unknown sex) and two cubs.[7] Sadly, almost

80 per cent of its habitat has been lost in a window of merely thirteen years (1970–83) according to a report by WWF.[8]

A DYNAMIC BEAUTY

With a distinct coat decorated with widely spaced rosettes bordered by dark steady rings, the Amur Leopard is one of the most stunning animals in the world. In addition to this elegant pattern, the pelt is functional and dynamic throughout the year. During summers, the hairs are shorter, about 2.50 cm., to keep the leopard cool; in the winter, they usually grow up to 7 cm. long to retain warmth. Moreover, as if it were replicating seasonal variations, the dynamic fur turns reddish-yellow colour during the warmer seasons and slowly turns into a lighter colour as the temperature falls.[9] Evolution has made it a beautiful, well-planned furry scarf out of its thick, bushy tail to survive the extreme winters in the Amur river basin (Heilong).

As the leopard reaches 2–3 years of age, they gain sexual maturity and start carving out their territory. Amur Leopards can live up to twenty years in captivity, but the average lifespan in the wild is unknown,[10] though other subspecies live up to 12–17 years.[11]

THE AMUR TERRITORY

Even though Amur Leopards are known to adapt to any habitat that provides it with sufficient food and cover, their favourite habitats are the mountain areas sheltered by forests, preferably with narrow watershed ridges and precipitous slopes up to heights of 300–500 m. above sea level. Its temperate forest habitat, where mercury can drop as low as -25° Celsius (-13° Fahrenheit), sets the Amur apart from other *Panthera Pardus*.[12] Thus, it has longer legs to navigate through the heavy snowfall and a distinct reddish-yellow winter coat, as mentioned earlier, with a golden tinge. It does not ascend higher altitudes, especially during winters.

A mature female leopard roams across an area of 35–45 sq. km.,[13] while a male needs an area four to six times larger. The vegetation in their habitat range covers cedar broadleaf, oak broadleaf, and black fir forests. Steady decline of habitats of Amur Leopards is a result of deforestation of coniferous species, ash trees, and mature oaks along with new road constructions that facilitate access to the most remote and inhospitable lands.[14]

The Amur Leopard's diet largely consist of roe deer (up to 66 per cent), wild boars (up to 8 per cent), Siberian musk deer (up to 9 per cent), sika deer (up to 6 per cent),[15] as well as other species such as weasels, badgers, birds, and mice during summers.[16]

THE SOLITARY NOCTURNAL

A crepuscular (nocturnal) animal, Amur Leopards can make use of low-light cover for hunting.[17] They are essentially solitary, nocturnal animals. However, according to certain reports, after mating some males stay with females, and may even assist in raising their young. Some males even follow and fight over a females.[18] However, the cubs are mostly reared by their mothers alone for the first two years of their lives. Despite being solitary in nature, leopards who have never met may recognize each other with scent markers, hunting trails or even migratory routes.[19]

Nimble-footed and robust, the leopard carries its unfinished kills and hides them so that they are not taken away by other predators.[20] It tries to get as close to its prey as possible before pouncing with extended arms and biting the back of the prey's neck.[21] After their nocturnal hunts, the leopards usually spend their daylight hours resting in caves or thickets.

Amur Leopards are *solitary prowlers*; aside from mating and territorial defence, they rarely intermingle with one another. However, the solitary animal has not been left with a solitary space due to incessant human greed and need, which are the root of almost all anthropocentric nature disasters.

LOSING GROUND: ANTHROPOCENTRIC THREATS

In a 2016 BBC report from the Shenfield High School, Essex, Erica, a student, said, 'I would feel sympathy for them as they have been skinned and people wear their fur.'[22] The child had sympathy for the beautiful animal being skinned, in sharp contrast to grown-ups (millions across the globe) who display pride on wearing an exclusive Amur skin outfit, or for that matter, any leather.

The leopard is at a feasibility threshold of extinction due to anthropogenic habitat alteration and increasing human populations, resulting in the leopard's habitat isolation, poaching, inbreeding, environmental randomness/stochasticity, and spread of infectious diseases.[23] The illegal wildlife trade is one of the biggest challenges being faced by the ever-dwindling leopard population. The Amur river basin is surrounded by villages and farmlands. Beauty can be exploitive at times and the gorgeous rosette-covered fur of the leopard has had a magnetic effect on poachers for a long time. The fur has a high demand in the luxury goods market. Isn't it worth a thought that 'fur is not fabric', and that we must change our perspective and understand that 'fur is beautiful on its owner and ugly on human'.

A 2016 report says that over 95 per cent of the original habitat of the leopard has been lost.[24] The species is in grave danger of extinction due to a number of factors, including habitat degradation (due to indiscriminate logging, heavy forest fires, and land conversion for farming), poaching and prey depletion, inbreeding and disease, and competitive interactions with tigers.[25] In the northern portion of the Changbai Mountains in Jilin and Heilongjiang provinces in China, bordering south-western Primorsky Krai in Russia to the east and North Korea to the south-west, the majority of forests have been logged, and many low-elevation forests have been transformed into secondary deciduous forests over the past five decades.[26] Many places have been transformed into open grasslands, which deer, and hence leopards, avoid.[27] Massive man-made fires, one of the major threats to Amur Leopards, have transformed forests into savannahs and grasslands, which are not appropriate for leopards. The purpose behind these fires is to burn the forests in a bid to stimulate the growth of certain types of ferns used in common Russian and Chinese dishes.[28]

Poaching is, by far, the biggest threat to the existence of the Amur leopard. Poachers hunt the leopards to trade in the wildlife market; their skin/fur is sold for a fortunes. Conscientious buyers can definitely play a huge part in wildlife conservation by saying 'No' to the fur/skin/leather market. Poachers do not even spare the leopards prey such as the Sika or Roe deer, for their meat can fetch them money. This has also resulted in a sharp decline in leopard prey. Currently, the prey population is not sufficient in Russia and China to sustain the leopards' numbers.[29] Additionally, ungulate poaching (using snares) pose a substantial conservation challenge to promote leopard retrieval.[30]

Sometimes even revival attempts, if not carried out in a scientifically appropriate way, can turn into misadventures. With a handful of leopards left in the wild, inbreeding and insufficient genetic miscellany can become a hazard. In such cases, disease and deformities are the resultant outcomes.[31] Also, there are a number of wild and domestic carnivores that co-exist with leopards and can be the carriers and transmitters for diseases.[32]

In another novel case in 2018, the *Journal of Wildlife Diseases* registered the first instance of canine distemper virus in a wild Amur Leopard, when a 2-year-old female was found on the side of the road demonstrating severe neurological symptoms, as reported in a paper.[33] Canine distemper virus (CDV) is a well-known disease in domestic dogs. It infects the respiratory system, causes intestinal problems and, in some cases, transmutes into to severe neurological issues. The mortality rate is high but differs depending on the species, and survivors mostly have painful lifelong side effects.[34] The study found that the 2-year-old leopard likely contracted the virus from proximate domestic dogs or wild carnivores, such as foxes or raccoon dogs, as there are very few leopards left to sustain the circulation of CDV themselves. While the virus may spread comparatively slowly between solitary species like leopards, even low rates of contagion critically increase the probability of extinction of the Amur Tigers[35] or ecologically similar solitary species such as the lion.[36] Making use of traditional conservation approaches like protecting their habitat, establishing surplus populations in another place/habitat and curbing the hunting of the big cats in general would help control and hopefully reverse their threated status.

REGENERATION ATTEMPTS

Nature knows no man-made boundaries. As such, humans should work as one unit to save its natural heritage, and the elusive Amur Leopard is one such charming aspect of nature. It is heartening to learn that China and Russia are developing a bilateral conservation strategy for the Amur Leopard to regrow its population. The Chinese government has come up with a multi-stage plan to expand the Amur leopard's territory and foray into the Changbaishan Mountain landscape. The process of restoring this fascinating flagship species in north-east Asian forest ecosystems will definitely catalyse broader conservation efforts in the region. However, China has a dearth of information about leopards, and as a comeback, the Chinese government is launching a science-driven approach to salvage the dwindling leopard population. Priority research initiatives comprise population estimations, monitoring the dispersal of leopards and identifying factors that threaten this subspecies so that human activities can be modified to foster leopard recovery.[37]

In December 2010, the announcement of the creation of a new national park—the 'Land of Leopard'—by Russian Deputy Prime Minister Sergei Ivanov was a landmark step towards the protection of this critically endangered species. The new, larger reserve planned to merge the Kedrovaya Pad Nature Reserve with the nearby Leopardovy Wildlife Refuge in Russia. Even Hunchun Nature Reserve in China, an important habitat for Amur Leopards, was slated to be added later to form a transboundary protected area.[38] The plan was finally implemented in 2012. Yury Darman, Director, WWF Russia's Amur Branch, observes, 'Even the first steps towards establishing the "Land of Leopard" national park are having positive results. The fact that the number of Amur leopards has grown from 7 to 12 on the monitoring plot offers proof that creating one united trans-boundary protected area is the right idea.'[39]

While it would be a shame to let a flagship species like the Amur Leopard go extinct, the very fact that their extinction will further destabilize an entire ecosystem also should be taken into account for further solidification of conservation strategies. Nature, as we know, is like an enormous clockwork apparatus and even slight meddling could lead to a colossal disaster. Amur Leopards are top predators in their territory, so they play a crucial role in keeping the balance of species in their area. For instance, without the leopard population, the deer, badger or hare population the area might explode, leading to over-grazing and other related perils.[40] This, in turn, affects the health of the forests and extended environment, which provides local wildlife and people in the adjacent areas with food, water and other resources.[41] Thus, saving the Amur leopard is an immediate need.

Despite the accomplishments of conservation organizations like the World Wildlife Fund (WWF), Wildlife Conservation Society (WCS) Russia, and the Amur Leopard and Tiger Alliance (ALTA), it is going to take people around the world to lend a helping hand to save this marvellous species. Not everyone can be an expert in wildlife or in planning strategies to curb poaching. But one can undeniably act as an aware and active citizen of the globe and play their parts, small or big, in redeeming what is left of this spectacular species. Let us all join hands to ward off the impending extinction of the Amur Leopard.

NOTES

1. Rebecca Greig, ed., *World of Animals: Book of Big Cats and African Wildlife*, Dorset: Future Publishing, 2018, p. 58.
2. 'Amur Leopard: Probably the World's Rarest Cat?', *WWF* (website), n.d., see https://www.wwf.org.uk/learn/wildlife/amur-leopards#:~:text=Adopt%20an%20Amur%20leopard%20today,be%20left%20in%20the%20wild, accessed 5 January 2023.
3. P. Jackson and K. Nowell, 'Panthera pardus ssp. orientalis. The IUCN Red List of Threatened Species', *Research Gate*, 2016, see https://www.researchgate.net/publication/308899520_Panthera_pardus_The_IUCN_Red_List_of_Threatened_Species_2016/citation/download, accessed 25 January 2023.
4. 'CITES Amendments', *CITES* (website), n.d., see https://cites.org/sites/default/files/eng/cop/08/prop/E08-Prop-EQ1_to_EQ5_Panthera.PDF, accessed 14 December 2020.
5. 'Amur Leopard', *WWF* (website), n.d., see https://wwf.panda.org/knowledge_hub/endangered_species/amur_leopard2/, accessed 14 December 2020.
6. L.M. Feng et al., 'First Image of an Amur Leopard Recorded in China', *Cat News*, vol. 55, no. 9, 2011.
7. Li Yang et al., 'Reconstructing the Historical Distribution of the Amur Leopard (Panthera pardus orientalis) in Northeast China based on Historical Records', ZooKeys, vol. 592, 2016, pp. 143–53.
8. 'Amur Leopard'.
9. 'Endangered Species', *WWF* (website), n.d., see https://wwf.panda.org/knowledge_hub/endangered_species/amur_leopard2/, accessed 14 December 2020.
10. 'Wildlife Ecology', *WCS Russia* (website), n.d., see https://russia.wcs.org/en-us/Wildlife/Amur-Leopards/Ecology.aspx., accessed 14 December 2020.
11. Greig, *World of Animals*.
12. Ibid.
13. 'Amur Leopard', *North-East Asian Subregional Programme for Environmental Cooperation* (website), n.d., see http://neaspec.org/sites/default/

files/Fact%20Sheet%20of%20Amur%20Leopard.pdf, accessed 14 December 2020.
14. Ibid.
15. Guwangshun Jiang et al., 'New Hope for the Survival of the Amur Leopard in China', *Scientific Report*, vol. 5, 2015.
16. D.G. Pikunov et al., 'A Survey of Far Eastern Leopards and Amur Tigers in Southwest Primorye Krai, in 2000', *WCS Russia* (website), 2000, see https://russia.wcs.org/en-us/Wildlife/Amur-Leopards/Ecology.aspx, accessed 14 December 2020.
17. Greig, *World of Animals*.
18. 'Amur Leopard: Species', *WWF* (website), n.d., see https://www.worldwildlife.org/species/amur-leopard, accessed 10 June 2023.
19. Greig, *World of Animals*.
20. 'Amur Leopard: Species'.
21. Greig, *World of Animals*.
22. Shenfield High School, n.d., see http://www.shenfield.essex.sch.uk/downloads/content/Endangered_Amur_leopards.pdf, accessed 14 December 2020.
23. D.G. Miquelle et. al., 'Identifying Ecological Corridors for Amur Tigers (Panthera tigris altaica) and Amur Leopards (Panthera pardus orientalis)', *Integr. Zool.*, vol. 10, 2015, pp. 389–402.
24. Andrew P. Jackonson et. al., 'Leopard (*Panthera pardus*) Status, Distribution, and the Research Efforts across its Range', *PeerJ*, vol. 4, 2016.
25. 'Wildlife Conservation Threats', *WWF* (website), n.d., see https://russia.wcs.org/en-us/Wildlife/Amur-Leopards/Conservation-Threats.aspx, accessed 14 December 2020.
26. T. Wang et al., 'A Science-based Approach to Guide Amur Leopard Recovery in China', *Biological Conservation*, vol. 210 (B), 2017, pp. 47–55.
27. Grieg, *World of Animals*.
28. 'Amur Leopard The Beautiful', *Luc Vertebrates* (blog), 2014, see https://lucvertebrates.wordpress.com/2016/12/09/amur-leopard-the-beautiful/, accessed 14 December 2014.
29. Grieg, *World of Animals*.
30. Y.H. Soh et.al., 'Spatial Correlates of Livestock Depredation by Amur Tigers in Hunchun, China: Relevance of Prey Density and Implications for Protected Area Management', *Biological Conservation*, vol. 169, 2014, pp. 117–27.
31. Grieg, *World of Animals*.
32. 'Wildlife Conservation Threats', *WCS Russia* (website), n.d., see https://russia.wcs.org/en-us/Wildlife/Amur-Leopards/

Conservation-Threats.aspx, accessed 14 December 2020.

33. Nadezhda Sulikhan et al., 'Canine Distemper Virus in a Wild Far Eastern Leopard (Panthera Pardus Orientalis), *Journal of Wildlife Diseases*, vol. 54, no. 1, 2018, pp. 170–4.
34. Melanie Greaver Cordova, 'Critically Endangered Amur Leopard Faces New Threat', *Cornell Chronicle*, 2018, see https://news.cornell.edu/stories/2018/01/critically-endangered-amur-leopard-faces-new-threat, accessed 14 December 2020.
35. M. Gilbert et al., 'Estimating the Potential Impact of Canine Distemper Virus on the Amur Tiger Population (Panthera tigris altaica) in Russia', *PLoS ONE*, vol. 9, no. 10, 2014.
36. Cordova, 'Critically Endangered Amur Leopard Faces New Threat'.
37. Wang, 'A Science-based Approach to Guide Amur Leopard Recovery in China'.
38. WWF, 'Critically Endangered Amur Leopards Captured on Video', *Phys Org*, 13 July 201, see https://phys.org/news/2011-07-critically-endangered-amur-leopards-captured.html, accessed 19 May 2020.
39. Ibid.
40. Greig, *World of Animals*.
41. 'Probably the World's Rarest Cat?', *WWF* (website), n.d., see https://www.wwf.org.uk/learn/wildlife/amur-leopards, accessed 14 December 2020.

PART II

ENDANGERED

10

Failing our Gods

An Elephantine Crisis

Prerna Singh Bindra

A DISTURBING PHOTOGRAPH by Biplab Hazra taken in 2015 in West Bengal, India, showing an elephant and her calf being hounded by a mob throwing *mashals* (rags set ablaze), became the face of the Human–Elephant Conflict (HEC) which has reached epic proportions across Asia and Africa because of damage to crops, property and reciprocal loss of life. In Hazra's photo, given below, you can hear the chaos, smell the burning tar, sense the fear, feel the torment. The mother, tail curled in terror, is on the run; followed by her calf, screaming in agony, its back enveloped in fire.

Unlike predators such as tigers or leopards, elephants do not prey on livestock; however, they can damage crops,

*Parts of this piece are extracted from the author's book, *The Vanishing: India's Wildlife Crisis*, Delhi: Penguin Random House, 2017.

The author is grateful to Ashok Kumar, Prakriti Srivastava, P.K. Sen, Paramjit Singh, Ramesh Pandey, Priya Davidar, Tarsh Thekaekara, Varun R. Goswami, M Ananda Kumar, Sanjay Pathak, Aditya Chandra Panda, Aritra Kshettry, Hannah Mumby, Sonia Jabbar, Divya Vasudev, and the many forest officers, guards, trackers and mahouts whose work has informed this article.

PLATE 10.1: A mother and calf try to escape the flaming tar bombs and crackers thrown by a mob in Bankura district, West Bengal, India, where depredation by elephants is severe. It has been learnt that the elephants escaped eventually. *Image courtesy*: Biplab Hazra

homesteads, and may occasionally kill people in accidental encounters. In India, which has nearly 60 per cent of the extant Asian Elephant (*Elephas maximus*) population—amounting to about 50,000—the conflict is severe and widespread. Over 400 people are killed annually by elephants. Though there has been no rigorous countrywide exercise to estimate the agricultural damage by elephants, it is believed to affect 8 to 10 million hectares of farmland annually.[1]

Such acute, consistent conflict depletes local support for conservation even in countries like India where elephants are revered and people are accepting of sharing spaces with wild animals.[2] In the long term, it can undermine and erode the persistence of elephant populations.[3]

While use of fireballs may be extreme, such reactionary, sometimes abusive, methods of driving and chasing elephants to keep them out of human habitation are almost routine in conflict-ridden hotspots. Conflicts with elephants date back thousands of years, as long as elephants and people have shared spaces; what has changed is their intensity and severity. Elephant habitats usually abut farmlands, which attract the animals. Crops offer easy power-packed nutrition. It is here, in this interface between farmland and forest, that the fault lines of human–elephant relations lie and where the tragic theatre of conflict is played.

Crop damage can be crippling. For instance, farmers around wildlife reserves in southern India may lose around 15 per cent of their annual produce to elephants.[4] I have met with farmers in Wayanad Wildlife Sanctuary in Kerala whose entire crop has been destroyed overnight by elephants.

So acute and vexing is the problem that at a private event (attended by the author in March 2019) Kenya's former environment minister Judi Wakhungu stated that that the threat to elephants from human-elephant conflict is growing to be more serious than poaching. This is shocking, considering about a hundred elephants are killed *everyday* globally by poachers—for their tusks, and increasingly, their meat.

The loss, stress, and suffering is colossal for people, and equally, for elephants, who are killed in retaliation to conflict—electrocuted, poisoned or killed by placing 'bombs' in food the animals forage on. A particularly poignant case in recent memory (June 2020) is from Kerala, where a pregnant elephant met with a slow, agonizing death when a crude bomb placed in the pineapple she was eating blew up.

ELEPHANT ECOLOGY

A key reason we are witnessing such accelerated conflict is our myopic view which fails to consider elephant ecology. Most conflict resolution is aimed to confine elephants, a wide-ranging species, to small spaces deemed 'protected areas'.

The matriarch is a repository of knowledge, and leads the herd. She has 'mapped' the physical as well as the social landscape: where foraging is possible, where water can be found in a pinch, which areas have more human disturbances, which herds are friendly and which are not.

Such information is crucial for the survival of the herd. When one such individual is killed, it has a knock-on effect on the herd. Without the knowledge and culture of the matriarch, the herd is rendered more vulnerable.

Elephants require 200 litres of water and 2 quintals of vegetable matter a day. The pachyderms, however, are not just copious consumers. Elephants are the gardeners of the forest as well. They are a crucial keystone species who shape ecosystems. Elephants disperse seeds as they travel, smartly packaged in fodder (i.e. dung)—sprouting new plants, and contributing, in a major way, to a forest's regeneration and plant diversity. Unlike other seed dispensers like birds or monkeys, who have smaller ranges, elephants travel vast distances, sometimes planting seeds far from the parent tree. In one study, scientists in the Democratic Republic of Congo found that forest elephants disperse about 345 seeds per day from 96 different species, typically more than 1 km. away from parent trees and shrubs, and sometimes, even as far as 54 km.[5]

Conserving such large, nomadic animals in India, now the most populous country in the world, is a challenge, especially given the intelligence of elephants.

An elephant's brain is comparable to humans in terms of structure and complexity. They are adept tool users. They cooperate and coordinate with each other to solve problems and work together for the greater good.[6] Forest officers in Palamu Tiger Reserve in northern India narrate how elephants 'test' the stability of flimsy wooden bridges across *nallahs* with their feet before they let the calves walk on them.[7] The adults, however, choose to swim across the *nallah*, which is dangerous for young calves in the turbulent waters during the monsoon.

One key strategy to mitigate conflict is minimizing interface between people and elephants by building barriers around protected areas. But elephant-proof trenches and electric fences may shift the conflict elsewhere, rather than solve the problem.[8] Barricading forests also fragment habitats and lead to population isolation. Besides, such fences are costly, poorly maintained, and the elephants inevitably figure out a way around them—by nudging mud into trenches to level them or pushing trees onto fences and walking over them!

The fear of, and antagonism towards, elephants has local people and foresters resorting to loud noises, bursting crackers, using pellets and even at times throwing burning metal rods or tyres to chase them away. These, too, fail to resolve the conflict, and can, in fact, be counterproductive. Use of such crude methods drastically increase stress levels in elephants, making them more aggressive and resulting in more damage during such conflicts. Such continued stress can affect overall elephant survival and reproduction.[9]

Typically, most measures dealing with HEC ignore the fact that biological, ecological and social causes are based on conditioning fear in elephants or are aimed at restricting and altering animal movement. They are largely unscientific and reactionary, and fail to address the underlying problems. Hannah Mumby and Joshua Plotnik, in a 2018 study, advocate an approach that 'takes the Elephants Perspective', integrating animal behaviour, cognition, ecology and knowledge of individual variation to address the underlying causes of HEC to prevent its occurrence.

THE THREATS: LOSS AND FRAGMENTATION OF HABITAT

In India, the elephant is god—Ganesha, the lord of good fortune, remover of obstacles, whose name is invoked before commencing any work or deed of significance, be it laying the foundation of a new home or setting up a business. In some parts of India, it is believed that after death, the human soul travels outside the body and moves into the souls of elephants.

Yet, even as we venerate Ganesha in temples, in the real world—in forests and fields, villages and towns—*Homo sapiens* are locked in conflict with the lord's earthly avatar.

The threat to elephants and other wildlife is driven by the overwhelming demand on natural resources. Relentless economic growth and consumption, intensifying land use for agriculture, and economic development have led to degradation, fragmentation and loss of wild habitats, pushing wild animals closer to humans. As conflict accelerates, peoples' tolerance to share space with the elephant is being tested.

'Linear infrastructure' such as highways and railway lines destroy forests, making long-range movement for animals difficult. Elephants are crushed by trains and electrocuted by illegal fences and low-hanging wires—about fifty are killed by electrocution annually in India.[10] Loss of forest cover directly causes and increases conflict. The elephant reserves in Sonitpur (Assam) and Kameng (Arunachal Pradesh) were massively encroached post the 1990s. By the year 2002, about half of the region's prime elephant habitat was lost, leading to a surge in human–elephant conflict. Between 1991 and 2003, ninety-three people were killed by wild elephants in this region. During the same time frame, fifty-two Elephants were found dead, twenty-five due to poisoning.

In north Bengal, elephants have lost nearly 70 per cent of their original habitat, taken over by tea estates, housing, fields, defence infrastructure, hydroelectric projects, highways, rail tracks, tourist resorts, etc. Pushed into human habitation, elephants kill about fifty people each year in the districts of Jalpaiguri and Darjeeling.[11] Elephants have been killed too; many bear scars and burns from lighted torches, and there are often bodies limp and pockmarked with bullets.

This conflict wounds the spirit as well. Calves born in, and living with, conflict are not unlike children raised in war zones. Anxious, traumatized, and having never known peace, they are more capricious, and prone to getting into conflict. Indeed, scientists say that constant stress is also a potential trigger for conflict.[12] Driving Elephants with searchlights, crackers or guns only makes them more aggressive. In an article, 'Elephant Breakdown',[13] Bradshaw et al. says that elephant populations are suffering from a form of chronic stress. Elephants are being poached for their ivory, poisoned, electrocuted and crushed by trains. They are watching their kin being slaughtered. The authors say that wild elephants are displaying symptoms associated with human Post Traumatic Stress Disorder—depression, unpredictable asocial behaviour, and hyper-aggression. While their article is based on elephants in Africa, this likely holds true for the Asian Elephant as well.

FAILING OUR GODS

In recognition of the deep cultural and religious significance, the elephant was declared India's National Heritage Animal in 2010. Established in 1992, 'Project Elephant' is dedicated to its conservation, but the initiative is underfunded and remains a poor cousin of Project Tiger or the National Tiger Conservation Authority, as it is now known. The elephant is listed under Schedule I of the Wildlife (Protection) Act, giving it the highest level of protection the law can offer. Paradoxically, the land the elephant inhabits is not protected: elephant reserves have no legal sanctity, with barely a third of them covered under the protected area network.[14] Elephant corridors—ancient migration paths that animals use to traverse from one forest to another—are unprotected as well, and are rapidly being blocked by mines, dams, roads, railways, canals, real estate and other construction.

Efforts to secure elephant habitats have consistently been resisted and thwarted by the powerful political–bureaucrat–business–mining nexus. In some ways, human–elephant conflict is a metaphor for the 'development vs. conservation' debate that is being played out in India, and indeed globally, today. The battle to protect the elephant has become Ground Zero for the war over natural resources. Elephant forests are storehouses of minerals. The Chotanagpur Plateau, spread across the states of Jharkhand and parts of Chhattisgarh and

Odisha, contains over 60 per cent of the country's iron ore.[15] These forests are being cleared to fuel the country's growth ambitions. In fact, India's former environment minister Jairam Ramesh cited coal and iron ore mining as the 'single biggest threat to Elephant habitats and corridors',[16] especially in the east-central Indian belt. In the battle between elephants and coal, the animals are expendable; tragically, so are the people caught in the crosshairs of the conflict.

The Indian government also failed to act on the recommendations of the Elephant Task Force it appointed in 2010. One key recommendation was to create an empowered National Elephant Conservation Authority on the lines of the National Tiger Conservation Authority, which has statutory powers under the federal Wildlife (Protection) Act, 1972.[17] The proposal never saw fruition, purportedly because it would be inconvenient for mining interests. At a cabinet meeting, a minister dismissed the proposed elephant authority, saying that 'there are enough elephants', so much so he 'could even see them on the streets', referring to the sorry sight of India's begging elephants—an outdated and cruel practice that allows domestication and trade of a protected wild animal.[18] Equally ludicrous is the myth of 'enough elephants'. As per official estimates India is home to about 27,000 Elephants, a 10 per cent decline from the last census in 2012.[19]

THE WAY FORWARD

With roughly 80 per cent of the current elephant distribution in India outside protected areas, the future of the elephant relies on peacefully sharing space with people as well as the latter's tolerance towards the animal.

There are no easy solutions, no magic spell to solve to this complex issue. However, there are some immediate must do's (cited here in no particular order): quick, fair and empathetic delivery of crop compensation and ex-gratia payments, awareness among local communities about personal safety, long-term work with local communities to gain their trust, and revival and use of traditional knowledge of elephant behaviour to minimize conflict. Installation of proper lighting and providing toilets facilities in conflict-ridden areas must be taken up, as many lives are lost in accidental encounters when people venture out in the dark.

Equally important is to work collaboratively with enforcement agencies like the police for crowd control—mobs exacerbate conflict. Driving elephants in human-dominated landscapes needs to focus on providing safe passage to people and elephants alike. During the harvest season, compensation amounts can be paid in advance to village councils for immediate disbursement to victims, rather than making farmers apply for compensation later, which is an exasperating process mired in bureaucratic delays.

There is a pressing need to look at approaches that address underlying issues, taking into account both human and wildlife factors. There are a few examples of successful use of such approaches. Human mortality and injuries due to leopards reduced in Mumbai when solutions were based on leopard ecology as well as awareness among affected communities. In HEC, such methods may be more relevant as elephants are cognitively complex, socially intelligent animals with distinct individual traits that may impact interaction with landscape and humans.[20] A long-term social awareness and public participation programme which includes an early warning system about elephant movement in Valparai, south India, reduced HEC intensity and showed marked increase in tolerance of people towards elephants as compared to other areas of conflict.[21]

Addressing the root cause of conflict—habitat destruction—is critical. Elephants are set to lose a further 42 per cent of their current habitat by the end of the century to anthropogenic pressure and climate change.[22] Where possible, protected areas can be expanded and land use in elephant habitats and corridors regulated. Allowing for safe passage of elephants through 'elephant corridors' is important. We need site-specific interventions, relevant to the place, habitat, culture, and problems.

It is the ethos of our people which accepts elephants and tigers in their landscape that has enabled their persistence in India, even as they have either become extinct or diminished in other countries where they formally ranged.

The need of the hour is concerted action to ease conflict and foster the veneration and the deep cultural links our people have towards elephants, indeed, towards all non-human life forms.

NOTES

1. S.S. Bist, 'An Overview of Elephant Conservation in India', *Indian Forester*, vol. 128, no. 2, 2002.
2. V. Athreya et al., 'Big Cats in our Backyards: Persistence of Large Carnivores in a Human Dominated Landscape in India', *PLoS One*, vol. 8, no. 3, 2013; P.S. Bindra, *The Vanishing: India's Wildlife Crisis*, Delhi: Penguin, 2017.
3. V.R. Goswami, V. Vasudev, M.K. and Oli, 'The Importance of Conflict-induced Mortality for Conservation Planning in Areas of Human–elephant Co-occurrence', *Biological Conservation*, vol. 176, 2014, pp. 191–8.
4. M.D. Madhusudan, 'Living amidst Large Wildlife: Livestock and Crop Depredation by Large Mammals in the Interior Villages of Bhadra Tiger Reserve, South India', *Environmental Management*, vol. 31, no. 4, 2003, pp. 466–75.
5. S. Blake et al., 'Forest Elephants: Tree Planters of the Congo', *Biotropica*, vol. 41, no. 4, 2009, pp.459–68; Jeremy Hance, 'Elephants: The Gardeners of Asia's and Africa's forests', *Mongabay* (website), 25 April 2011, see https://news.mongabay.com/2011/04/elephants-the-gardeners-of-asias-and-africas-forests/, accessed 15 December 2022.
6. Ferris Jabr, 'The Science is In: Elephants are Even Smarter Than We Realized', *Scientific American*, 26 February 2014; Joshua M. Plotnik et al., 'Elephants Know When They Need a Helping Trunk in a Cooperative Task', *Proceedings of the National Academy of Sciences*, vol. 108, no. 12, 2011, pp. 5116–21.
7. Personal Interview.
8. L. Osipova et al., 'Fencing Solves Human-wildlife Conflict Locally

but Shifts Problems Elsewhere: A Case Study using Functional Connectivity Modelling of the African Elephant', *Journal of Applied Ecology*, vol. 55, no. 6, 2018, pp. 2673–84.

9. H.S. Mumby and J.M. Plotnik, 'Taking the Elephants' Perspective: Remembering Elephant Behavior, Cognition and Ecology in Human-elephant Conflict Mitigation', *Frontiers in Ecology and Evolution*, vol. 6, 2018, p. 122.
10. Chiranjeevi Kulkarni, '600 elephants electrocuted in India in 10 years', *Deccan Herald*, 17 December 2021, see https://www.deccanherald.com/state/600-elephants-electrocuted-in-india-in-10-years-1061688.html, accessed 15 December 2022.
11. 'Radio–Telemetry Study of Elephants in Buxa Tiger Reserve and Adjoining Areas in North West Bengal', Asian Nature Conservation Foundation (website), 2013, see http://www.asiannature.org/our-programmes/asian-elephant-research-and-conservation-aerc/human-elephant-conflict-mitigation-pr-2, accessed 15 December 2022; S. Chakraborty, 'Human-animal Conflicts in Northern West Bengal: Losses on Both Sides', *Int J Pure App Biosci*, vol. 3, 2015, pp. 35–44.
12. G.A. Bradshaw et al., 'Elephant Breakdown', *Nature*, vol. 433, no. 7028, 2005, p. 807.
13. Ibid.
14. M. Rangarajan et al., *Gajah: Securing the Future for Elephants in India*, Report of the Elephant Task Force, Delhi: Ministry of Environment and Forests, 2010.
15. Kisor Chaudhuri, 'Impact of Mining on Elephants of Chhotanagpur Plateau, Central India', *Gajah*, vol. 42, 2015, pp. 30–3.
16. WTI, 'Shared Concerns for Elephants Bring Together African and Asian countries at the E-8 Meet', Wildlife Conservation, 24 May 2011, see retrieved from https://www.conservationindia.org/news/shared-concerns-for-elephants-bring-together-african-and-asian-countries-at-the-e-8-meet, accessed 15 December 2022.
17. Rangarajan et al., *Gajah*.
18. Bindra, *The Vanishing*.
19. Mayank Aggarwal, 'India's Elephant Population Decreases by 10% to 27,312', *Live Mint*, 12 August 2017, see https://www.livemint.com/Politics/OhPuqM871gkivaj5zetCZJ/Indias-elephant-population-decreases-by-10-to-27312.html, accessed 15 December 2022.
20. Mumby and Plotnik, 'Taking the Elephants' Perspective'.
21. Tarsh Thekaekara, 'Can Elephants and Humans Live Together?', *The Guardian*, 6 March 2017, see https://www.theguardian.com/environment/2017/mar/06/can-elephants-and-humans-live-

together, accessed 15 December 2022; Anandakumar Mavatur and Ganesh Raghunathan, 'Fostering Human–elephant Coexistence in the Valparai Landscape, Anamalai Tiger Reserve, Tamil Nadu', in *Human-Wildlife Conflict in the Mountains of SAARC Region Compilation of Successful Management Strategies and Practices*, Thimpu: SAARC Forestry Cente, 2014.

22. R. Kanagaraj et al., 'Predicting Range Shifts of Asian Elephants under Global Change', *Diversity and Distributions*, vol. 25, no. 5, 2019, pp. 822–38.

11

Red Panda
The Solitary Mountainous Mammal

Sunita Pradhan

> I think of what wild animals are in our imaginations. And how they are disappearing — not just from the wild, but from people's everyday lives, replaced by images of themselves in print and on screen . . . .
>
> —Helen MacDonald

The red panda is a shy, solitary mammal of the mountains. Very few will deny the charm of a Red Panda, be it in a zoo or in the wild. It is a very charismatic animal that stands out even more when it is seen in its habitat, the mountain forests. With cool climate and breeze, tall trees of birch, evergreen conifers, Rhododendrons with mosses and lichens hanging from them, leafy lanky bamboo, and glistening brooks—the Red Panda's habitat is as beautiful as the animal itself.

Conservation Status

The Red Panda is an endangered species according to the IUCN Red Data List. The species is protected in all its range countries. In India, the Red Panda is listed as a Schedule I species in the Indian Wildlife Protection Act (1972).

PLATE 11.1: The Red Panda
Image courtesy: Dhrithiman Mukherjee

RED PANDA: THE LESSER PANDA

There is yet another panda, the Giant Panda, which is even more popular and well known than the Red Panda. The Giant Panda is a species native to central China and has long been a logo of the World Wildlife Fund (WWF). As the Red Panda is smaller in size than a Giant Panda, the former it is also often called a 'Lesser Panda'. However, the Red Panda was known to the world outside their native habitat earlier than the Giant Panda. The Red Panda was first described by Major General Thomas Hardwick in 1821, but his paper was published until six years later. During this time, Fedrick Cuvier, a French naturalist, in 1825, published the description of the Red Panda and gave it its scientific name, *Ailurus fulgens*. Outside its native region of south China, the Giant Panda became known to the world around 1869.

The Red Panda and the Giant Panda have a few similarities, especially their specialized diet of bamboo, similar habitat, and physical characteristics such as thumbs. Other than these few similarities, the two species of panda are not related. The Red Panda is closer to a group of mammals known as Mustelids, which consists of animals like weasels, otters, racoons. Despite its relations with other mammal groups, the Red Panda is distinct, unique and is in a Family of its own, known as *Ailuridae*.

There are two species of Red Panda, something which was recently confirmed[1] by extensive molecular studies. The two species are *Ailurus fulgens* or the Himalayan Panda and *Ailurus styni* or the Chinese Panda. Scientists further report that the two species of pandas diverged from each other 250,000 years ago.[2] The Red Panda's ancestor can be traced back to tens of millions of years ago with a wide distribution across Eurasia.[3] However, the Red Panda we know and see today is found only in Asia, specifically in five Asian countries.

DISTRIBUTION OF THE RED PANDA

The five countries with Red Panda distribution are Nepal, India, Bhutan, Myanmar and China. The habitats of Red Panda within these countries lie within an altitude range of 2,200 m. to 4,000 m., covering an area of almost 134,975 km^2 in its distribution range.[4] The Himalayan Panda is found in Nepal, north-eastern India, Bhutan, Myanmar, Tibet and western Yunnan province of China, while the Chinese Red Panda is found in India and the Yunnan and Sichuan provinces of China.[5] In India, the Himalayan Panda is found in the state of Sikkim as well as Darjeeling and Kalimpong districts of West Bengal, while Arunachal Pradesh houses both the species with Siang River being the boundary separating these two phylogenetic species.[6] Although India has only 5 per cent of the total Red Panda habitat, it has a significant proportion (56.80 per cent) of the most suitable Red Panda habitat.[7] As many as twenty-two protected areas hold Red Pandas in India. One of the earliest studies on the Red Panda for its conservation in India was done in the Singhlaila National Park in Darjeeling, West Bengal.

RED PANDA: THE CAT BEAR

The Red Panda is also known as a cat bear, a name which convey the animal's appearance. The Red Panda is slightly bigger than an average domestic cat, measuring about 60 cm. in body length with a tail about 40 cm. long. Adults in the wild weigh 4 kg. on average and in captivity they weigh 4–5 kg.[8] The head is round with a short muzzle and pointed and erect ears, tufted with white hairs. The face is predominantly white, with vertical reddish-brown stripes below the eye which extend to the corner of the mouth. The body is heavily built and the legs are short and stump-like. The pelage of Red Panda is composed of long, coarse hairs to keep the animal warm and dry in its cold and moist environment. The upper body pelage is chestnut brown, while the lower side of the body and the limbs are black.

The colouration of the Red Panda perfectly camouflages the animal in branches of fir trees with thick cushion of reddish-brown moss. The bushy tail has twelve alternate rings of red and buff coloured, long hairs. The soles of the feet are covered by a dense, white mat of hair. The well-developed five toes on each foot are widely separated. The curved claws are semi-retractile, which means that the claws remain protruded, while in most cats the claws are retractile, i.e. they are withdrawn into a sheath on its paws. The claws are drawn out while hunting, running, or climbing. Their claws and our nails are made of the same substance known as keratin.[9]

The Red Panda has a plantigrade gait, which means it walks on its soles, like human beings and a few other animals like the bear. The Red Panda can climb trees, aided by its sharp curved claws. In fact, it is an arboreal animal, spending most of its time on trees. The strongly developed musculature in its forelimbs and in its tail helps the animal to adeptly walk on tree branches while balancing its body to prevent falling off. It descends trees head first, something that very few animals do. Its vocal repertoire includes squeals, twitters, and quack-snorts; non-vocal sounds include hisses, jaw clapping, and grunts.[10]

The thumbs of a Red Panda are one of its significant features. It is also known as a pseudo thumb. The pseudo thumb is a common feature between the Red and the Giant Panda.

RED PANDA AND ITS BAMBOO DIET: AN UNUSUAL CARNIVORE

The Red Panda's pseudo thumb is associated with its diet. It is a carnivore; it has sharp canine teeth and a short gut or[11] alimentary canal. However, even with all the characteristic features of a carnivore, the Red Panda, interestingly is an unusual carnivore; it does not have a meat-based diet and, instead, subsists predominantly on a leafy diet of bamboo leaves. The pseudo thumb is specially developed to grip bamboo clumps while eating the leaves. Although bamboo forms 90 per cent of Red Panda's diet, they also eat fruits and mushrooms. Bamboos are also an unusual plant as most of them flower only every 60 to 130 years depending on the species. Once they flower the plant dies, as it uses a large amount of energy to produce flowers and seeds.[12] Bamboo has a strong bearing on Red Panda ecology and survival.

HABITAT

Red Pandas live in habitats within an altitude range of 2,200–4,000 m., which covers the temperate and sub-alpine forests, with thick bamboo understorey. As the Red Panda is an arboreal mammal, trees are an important component of its habitat.[13] Trees are almost like a playground for the Red Panda; we have seen cubs play around on trees.[14] Hence a good cover of trees, bamboo, and suitable distance to water sources are important constituents of a Red Panda habitat.

BIRTH AND CUBS

A female Red Panda gives birth in the monsoons around June–July, when the forest has abundant food. The litter consists of an average of two cubs, after a gestation period of approximately 135 days. In the wild, Red Pandas gives birth in hollow trees. Mother pandas have been seen leaving their cubs for as long as eight hours when going out to feed. This is when the cub is all by itself, vulnerable to any kind of attack by animals and human beings.[15] The cubs are very tiny at the time of birth.[16]

HOW IS THE RED PANDA THREATENED?

The Red Panda is 'Endangered' because its population has possibly declined by 50 per cent over the last eighteen years; scientists are afraid that this rate of decline could continue.[17] Some of the threats and challenges to the Red Panda[18] are:

1. Bamboos dying off: Red Pandas have a staple diet of bamboo which die off after blooming, leaving the species to find new feeding grounds in a highly fragmented landscape. The Red Panda has shown poor survival in fragmented areas. These bamboos do not easily re-establish themselves after flowering in areas of environmental degradation and deforestation, which are now widespread across the species' range.
2. Diseases: Red Pandas are highly susceptible to canine distemper (even developing the disease after vaccination with domestic dog vaccine), which is lethal to them.
3. Habitat specialization: The Red Panda has specific habitat

requirements and will not survive if these requirements are not met. However, the habitat of Red Pandas is being fragmented and lost due to various land-use changes. Species with specialized needs of habitat and food will be the ones who will suffer most in face of such rampant habitat loss.
4. Illegal trade: Hunting for pet trade is increasing for Red Panda more than ever before.

WHAT HAPPENS IF THE RED PANDA GOES EXTINCT?

The Red Panda is not only a beautiful species but is also an animal with all the characteristics of a carnivore; yet it is completely dependent on a specialized diet of bamboo leaves. Therefore, losing such a unique herbivorous–carnivorous species is a great loss to the diversity that nature has to offer.

WHAT CAN WE DO FOR RED PANDA CONSERVATION?

The Red Panda's survival in the wild is challenged by destruction of its habitat. Tourist visits to national parks which are homes to the Red Panda have been a major concern. To cater to the demand of tourists, the habitats of Red Pandas are exploited and destroyed. Hence, it is important to be responsible while visiting national parks which hold special, often endangered, animals.

Working towards stopping wildlife trade and towards decreasing the demand of wildlife and their body parts is of utmost importance. Red Panda populations grow very slowly, as females give birth to very few cubs in a year, which are very tiny and vulnerable. Moreover, it takes almost eighteen months to reach an age when the female panda can give birth. Considering all these factors, Red Panda populations grow very slowly in the wild. And on top of this, if Red Pandas are rampantly hunted and killed, it becomes very difficult for the population to recover in the wild.

Domesticated dogs can also be dangerous for the species not only because they chase and attack Red Pandas but also because they transmit diseases to the wild species. Hence, vaccination of domestic dogs in fringe villages is very important. At the same time, to save Red Pandas and all other organisms in the world, the consumption pattern of people must be mindful and sustainable.

Research studies on wild species are important to understand their ecology and their responses to various threats such as climate change. The Red Panda can serve as a critical indicator species to monitor the health of the Himalayan ecosystem. Let us then pledge to save the beautiful Red Panda.

NOTES

1. Y. Hu et al., 'Genomic Evidence for Two Phylogenetic Species and Long-term Population Bottlenecks in Red Pandas', *Science Advances*, vol. 6, no. 9, 2020.
2. Ibid.
3. E. Mayr, 'Uncertainty in Science: Is the Giant Panda a Bear or a Raccoon?', *Nature*, vol. 323, 1989, pp. 769–71.
4. A. Thapa et al., 'Predicting the Potential Distribution of the Endangered Red Panda across its Entire Range using MaxEnt odelling', *Ecology and Evolution*, vol. 8, 2018, pp. 10552–4.
5. Hu et al., 'Genomic Evidence for Two Phylogenetic Species'; B.D. Joshi et al., 'Siang River in Arunachal Pradesh Splits Red Panda into Two Phylogenetic Species', *Mammalian Biology*, vol. 101, no. 1, 2021, pp. 121–4.
6. Joshi et al., 'Siang River in Arunachal Pradesh Splits Red Panda'.
7. Thapa et al., 'Predicting the Potential Distribution of the Endangered Red Panda'.
8. M.S. Roberts and J.L. Gittleman, 'Ailurus fulgens', *Mammalian Species*, vol. 222, 1984.
9. T. Safaris, 'Claws Out: Things You Dont Know about Claws', *Thomas Safaris* (blog), 2014, see https://thomsonsafaris.com/blog/animal-claws-out/, accessed 20 June 2020.
10. Mayr, 'Uncertainty in Science'.
11. M.L. Roberts, 'On the Subspecies of Red Panda Ailurus fulgens', *The Red or Lesser Panda Studbook*, vol. 2, Rotterdam: Stichting Koninklijke Rotterdamse Diergaarde, 1982, pp. 13–24.

12. 'Bamboo Flowering', *Bamboo Plants and Anatomy* (website), 2019 see http://www.bamboobotanicals.ca/html/about-bamboo/bamboo-flowering.html, accessed 13 June 2020.
13. Roberts, 'On the Subspecies of Red Panda Ailurus fulgens'.
14. S. Pradhan, 'Some Aspects of Ecology of the Red Panda, Ailurus fulgens in the Singhalila National Park, Darjeeling, India', PhD dissertation, North Bengal University, 1999.
15. P.B. Yonzon and M.L. Hunter Jr., 'Conservation of the Red Panda Ailurus Fulgens', *Biological Conservation*, vol. 57, 1991, pp. 1–11.
16. M.S. Roberts and J.L. Gittleman, 'Ailurus fulgens', *Mammalian Species*, vol. 222, 1984.
17. A. Glatston et al., 'Ailurus fulgens (errata version published in 2017)', The IUCN Red List of Threatened Species (website), 2015, see https://dx.doi.org/10.2305/IUCN.UK.2015-4.RLTS.T714A45195924.en, accessed 4 February 2022.
18. Ibid.

12

The Endangered Bovine
Wild Water Buffalo

Anwaruddin Choudhury

> We all have a responsibility to protect endangered species, both for their sake and the sake of our future generations.
>
> —Loretta Lynch

The domesticated buffalo is a familiar animal across the globe. In countries like India, Pakistan, Thailand, Brazil and China, it is an important part of the social and economic

*The author acknowledges Professor Mohd. Taher, Nagen Sharma, Pradyut Bordoloi, Dr Anil Kr. Goswami, L. Rynjah, Dr D.K. Sharma, M.C. Malakar, B.S. Bonal, Anindya Swargiary, P.M. Lad, Colin P. Groves, J.W. Duckworth, Paniram Das, Nur Hussain, Bikul Goswami, Padma Dehingia, Humon Mia, Lankeswar Pegu, Bolo Paul, Dilwar Hussain, Mukunda Hazarika, Manuranjan Baruah, Bhubaneswar Sharma, Mohabbat Shah Ali, Imran (Babu), Mukut, Raju, Hakim, Hashim, Leto Mili, A.M.K. Bharos, Kishore Rithe, F. Sap Bawia, Thomas Kent, Khekiho Sohe, Heerak Nandy, Romu Mazumder, Sajjad Choudhury. Kamal Kalita, Dilip Handique, late Sakul Boro, Babul Debnath, Faizul Ali, Moniram Boro, C.R. Bhobora, Narayan Sarmah, Dharanidhar Boro, Labanya Ramchiary, Dipen (Raju) Sonowal, Abdul Majid Choudhury, Abdul Munim Majumdar, Alauddin Choudhury, Hena Choudhury, Bilkis Begum and other well-wishers who gave me the benefit of their advice.

lives of millions of people. However, a lesser-known fact is that the buffalo's ancestor is the rare Wild Water Buffalo (*Bubalus arnee*) from Assam and some pockets of South and South-East Asia. Published information on the species is relatively scarce, even though it is listed by IUCN (2023) as an 'Endangered' species. The Wild Water Buffalo is rare with a small and fragmented range; the species has already vanished from a majority of its former range.

Perhaps the earliest documentations on Wild Water Buffalo can be found in *The Babur-nama* (*Memoirs of Babur*) in the sixteenth century.[1] After a long gap of around 400 years, A.A. Kinloch,[2] W.S.J. Richardson,[3] J.D. Inverarity,[4] and G.P. Sanderson[5] provided valuable information on the species. Thereafter several articles in the *Journal of Bombay Natural History Society* provided some more information. The first detailed account on the distribution and status of the Wild Water Buffalo along with mapping for north-east India was done by A.U. Choudhury.[6] Since 1990s, status reports on several areas of north-eastern India were published[7] to study the population genetics of Wild Water Buffaloes in Assam's Kaziranga National Park, alongside a few articles on its behavioural aspects.[8] A comprehensive documentation in the form of a monograph was produced in 2010[9] with an update in 2014[10] by Choudhury.

RANGE, HABITAT AND POPULATION

The historical range of the Wild Water Buffalo is somewhat uncertain due to at least 5,000 years of domestication in India, China and other countries. However, the inferred historical range extended from northern India to Indo-china and perhaps up to the Euphrates River in Iraq in the west.[11] The more recent range of the Wild Water Buffalo extended from the north Indian *terai* to southern Vietnam, with the river Godavari being the southern limit in peninsular India.[12] By the turn of the twentieth century, the animals remained in three large disjunct areas—northern India including Nepal,

Bhutan and Bangladesh, peninsular India, and scattered areas in Indo-china.[13] The species has indisputably have become extinct in large parts of its former range and now only highly fragmented populations occur.

The current distribution of the Wild Water Buffalo is confined to a few pockets in north-east and central India (Assam, Arunachal Pradesh, Meghalaya, West Bengal, Chhattisgarh and Maharashtra) and small pockets in Bhutan, Nepal, Thailand Cambodia, and probably Myanmar. The species has been extirpated from Bangladesh and Lao PDR, and its population doesn't seem to be rising in Thailand and Cambodia, as the reports indicate. The small number of wild buffaloes of Meghalaya is also nearing extinction.

The state of Assam is the stronghold of the Wild Water Buffalo, with four-fifths of the known population. As a single site, Kaziranga National Park has the largest population of the species globally. Some other key sites in Assam are Manas and Dibru-Saikhowa National Parks, and Laokhowa and Burhachapori Wildlife Sanctuaries. In central India, the main strongholds for this species are Indravati National Park, Pamed and Udanti Wildlife Sanctuaries (all in Chhattisgarh) and Kolamarka Conservation Reserve (in Maharashtra) Outside India, the key sites for the species are Koshi Tappu Wildlife Reserve in Nepal and Huai Kha Khaeng Sanctuary in Thailand.[14] There are also past records of the species from the Barak valley districts of Assam and from parts of Manipur, Mizoram, Nagaland and Tripura. Stray individuals are occasionally recorded in Odisha.

The Wild Water Buffalo's habitat is strongly linked to tall grasslands with water bodies, preferably in low-lying areas. The animals are both diurnal and nocturnal. In disturbed habitats, they are normally nocturnal. Wallowing is an important activity in a Wild Water Buffalo's daily life. The females of a herd often keep their young ones in one open area and move away to graze (as observed in Kaziranga). They may also leave their calves with other adult/adults and move a few hundred metres away. In peninsular India, the

PLATE 12.1: A wild buffalo in Kaziranga National Park, where the largest population of the species occurs.
Image courtesy: Anwaruddin Choudhury

buffalo frequents grassy glades and open grasslands amidst woodlands in the morning and evening, spending the hot hours of the day inside forests or in pools.

In 1948, S.H. Prater[15] mentioned that buffaloes have a keen sense of smell and good hearing; however, their sight is moderate. Adult bulls often fight during the mating season for dominance. As a rule, the Wild Water Buffalo is not very shy, and in areas such as Kaziranga National Park, with its long history of protection and lack of poaching, they are easily seen and can be approached up to a reasonable distance (within 5 m. on elephant-back). However, in areas where they are hunted for meat or where they cannot move freely during the daytime owing to human activities, they are wary and shy. Although buffaloes are mainly grazers, they also feed on herbs, fruits, and bark as well as on trees and shrubs. Hence, a mosaic of habitats with some grassy areas can also house the animal. Wild Water Buffaloes going out to feed on cultivated crops is a common phenomenon near the fringe areas of their habitat.

Wallowing in pools, rivers and mud is an important activity in the animal's daily life. It is basically part of its 'rest' period and may continue for a few hours. To wallow, the animals resort to using small jungle pools, the edges of large and small *beels*, marshy areas, rivers and fast-flowing streams. The apparent function of wallowing is to have a cooling effect as buffaloes normally do not wallow on cool winter mornings and at nights. According to some authorities, wallowing has a thermoregulatory function, and mud prolongs the cooling effect by slowing evaporation.[16] Studies on feral buffaloes in Australia have shown that wallowing may not be essential for the survival of the buffalo as, when deprived of the water bodies to wallow in, the animals behaved like cattle and seek shade.[17] An additional and conspicuous advantage of wallowing is the thick coat of mud on the body, which protects the buffaloes from insect bites, even if for a short time.

Crop raiding by Wild Water Buffaloes is a common phenomenon near their habitat while lone bulls mate with domestic cows in fringe areas. Buffaloes are aggressive in their behaviour; in particular, solitary bulls and cows with calves are dangerous and may charge people. This fact was mentioned as far back as the sixteenth century by Zahiruddin Muhammad Babur (*c.*1526)[18] in his *Babur-nama*.[19] Inverarity,[20] by contrast, mentioned that 'A herd of buffalo never charge, sometimes they bolt.'

A detailed population estimate is given by Choudhury,[21] which puts the number at about 3,400–3,500, which has shown some improvement owing to rise in numbers in Assam and Nepal. The current estimate is around 4,600–4,800 (Table 12.1). In Manas National Park (Assam), the population increased from 230 in 2008 to 500+ currently. D'Ering Memorial Wildlife Sanctuary, Arunachal Pradesh, has also seen an increase owing to improvement in habitat and reduction in poaching but elsewhere in the same state there is a decline in population. In Nepal's Koshi Tappu Wildlife Reserve, there is significant increase mainly due to lack of any natural predation. The tiger has been extirpated, and the leopard is also very rare, which, however, is generally not a threat to adult animals.

TABLE 12.1: Estimates of the Number of Wild Water Buffaloes

Country	Province/State	Number of Wild Buffalo*	
		In Choudhury (2010)	Current estimate
India	Assam	2,800	3,800
	Arunachal Pradesh	200	150
	Meghalaya	50	<40
	Other States	<50	>70
	India Total	3,100	4,070
Nepal		200	510
Thailand		50	50
Cambodia		<20	<20
Bhutan		<30	<40
	Rest of the World Total	300	620
TOTAL		**3,400**	**4,690** (4,600–4,800)

Source: Updated from Choudhury (2010)
*91.20 per cent in India; 82.30 per cent in Assam; 57.70 per cent in Kaziranga (Choudhury 2010); 86.80 per cent in India; 81 per cent in Assam; 53.50 pe cent in Kaziranga (this study).

CONSERVATION

Buffalo populations have experienced a severe decline due to habitat destruction, poaching for meat, competition and disease transmission from domestic stock, possible hybridization in some localized area, insurgency and invasion of exotic weeds.

The Wild Water Buffalo is still threatened with extinction across its range as the population is still not large enough and its habitat has remained the same, with little possibility of increase. Habitat loss has already restricted the species to a few pockets. Its typical habitat is grassland in fertile plains, which is the prime land for wet paddy cultivation and human habitation. As a result, such habitats have quickly vanished during the last century. Currently most of the habitat is found inside protected areas. Poaching for meat had resulted in loss of nearly four-fifths of the buffalo population in Manas National Park of Assam in the 1990s and early years of the twenty-first century. While the situation in Manas has now improved, the issue still persists in Dibru-Saikhowa National Park.

In the Indian subcontinent, hunting and poaching of Wild Water Buffaloes is mainly for meat. The earliest information on hunting the species is found in the *Ain-i-Akbari* by Abu'l Fazl (1551–1602).[22] According to him, during the reign of Akbar, the commonest method of hunting Wild Water Buffalo was to use a tame female in heat as a decoy. When the wild bull approached to mate, it was caught by hunters

PLATE 12.2: A herd of wild water buffaloes in Manas National Park
Image courtesy: Anwaruddin Choudhury

who slipped nooses around his legs. The memoirs of the maharajah of Cooch Behar[23] showed that the hunting pressure, for hobby, pastime and trophy by the royal dignitaries was also significant. Between 1871 and 1907, his royal hunting parties killed at least 434 Wild Water Buffaloes in a relatively small area comprising present-day northern West Bengal and western Assam. The number of injured animals would presumably have been more than that. The maharaja's hunting parties also once included Lord Curzon, the Viceroy of British India.

Dev Goswami[24] mentions that even in the 1930s, there was rampant poaching in Majuli and Lakhimpur in Assam. Although no systematic record is available, every year an unspecified number of buffaloes are shot outside protected areas, mostly in central India and in the North-East. Most such killings go unreported.

Outbreak of various contagious diseases spread by domestic livestock such as rinderpest, foot-and-mouth (FMD), and anthrax are also threats to the Wild Water Buffalo. Outside protected areas, hundreds of buffaloes and gaurs died from time to time; however, most of these deaths went unrecorded. Fortunately, rinderpest has been eradicated from India to a great extent, with no record of any death (either domestic or wild animals) for over two decades.

Mining is a serious issue in central India and there are oil wells near buffalo habitats in Dibrugarh and Tinsukia districts, Assam, especially near Dibru-Saikhowa National Park.

More than half the Wild Water Buffalo populations are found inside a single protected area (Kaziranga National Park, Assam). So far, more than ten separate populations/ sub-populations have already been formed in what were once largely contiguous populations. Many of these are non-viable and susceptible to genetic degradation.

The government of Chhattisgarh has declared the Wild Water Buffalo as its state animal. At least thirteen national parks and sanctuaries across its range protect the species. It is also protected by law in most of its range countries. In India, it is listed in Schedule I of The Wild Life (Protection) Act 1972, which prohibits its killing or capture. This is, in fact, the highest level of protection accorded to any species in the country. In 2017, Nepal translocated twelve animals from Koshi Tappu and three from Kathmandu Zoo to Chitwan National Park. This number is now eleven, including two calves born at Chitwan. At least six lost their lives due to disease and tiger predation.

It is suggested that steps need to be taken for effective protection and management of existing protected areas: fencing with metal railings in some areas, control of poaching, and involvement of local communities in conservation. Further, detailed surveys in Myanmar, Lao PDR and Vietnam, population estimate and monitoring at regular intervals and genetic studies are the need of the hour. Removal of exotic weeds, reintroduction in select former habitats (starting with Dudhwa National Park in Uttar Pradesh, India and in some reserves in Nepal), checks on encroachment, eviction of encroachers, creation of new protected areas, extension of existing protected areas and conservation education could also help save and nurture the species.

NOTES

1. S. Ali, 'The Moghul Emperors of India as Naturalists and Sportsmen. Part II', *Journal of the Bombay Natural History Society*, vol. 32, no. 1, 1927, pp. 34–63.
2. A.A. Kinloch, *Large Game Shooting in Thibet, the Himalaya, Northern and Central India*, Calcutta: Thacker, Spink & Co., 1885; A.A. Kinloch, *Large Game Shooting in Thibet, the Himalaya, Northern and Central India*, 3rd revd. and enlarged edn., Calcutta: Thacker, Spink & Co., 1892, p. 307.
3. W.S.J. Richardson, 'Correspondence: Wild Buffalo', *Journal of the Bombay Natural History Society*, vol. 4, no. 3, 1889, p. 231.
4. J.D. Inverarity, 'The Indian Wild Buffalo', *Journal of the Bombay Natural History Society*, vol. 10, 1895, pp. 41–5.
5. G.P. Sanderson, *Thirteen Years Among the Wild Beasts of India*, 1879; repr., London: W.H. Allen & Co., 1896, p. 387.

6. A.U. Choudhury, 'The Decline of Wild Water Buffalo in Northeastern India', *Oryx*, vol. 28, no. 1, 1994a, pp. 70–3.
7. A.U. Choudhury, 'Saving the Wild Buffalo', *The India Magazine*, September 1993, p. 75; A.U. Choudhury, 'Wild buffalo *Bubalus bubalis* in Dhakuakhana, Lakhimpur district of Assam', *Journal of the Bombay Natural History Society*, vol. 91, no. 1, 1994, pp. 137–8; A.U. Choudhury, 'Wild Water Buffalo *Bubalus bubalis/arnee* in Dibang Valley District of Arunachal Pradesh', *Journal of the Bombay Natural History Society*, vol. 95, no. 1, 1998, pp. 110–12; A.U. Choudhury, 'Status of the Wild Water Buffalo *Bubalus arnee* in Lohit District, Arunachal Pradesh', *Journal of the Bombay Natural History Society*, vol. 96, no. 1, 1999, pp. 130–2; A.U. Choudhury, 'On the Occurrence of the Wild Water Buffalo *Bubalus arnee* in the Barak Valley Districts of Assam', *Journal of the Bombay Natural History Society*, vol. 98, no. 2, 2001, pp. 270–1; A.U. Choudhury, 'Occurrence of the Wild Water Buffalo *Bubalus arnee* in Mizoram and Manipur', *Journal of the Bombay Natural History Society*, vol. 99, no. 3, 2002, pp. 512–16.
8. K.R. Ashby and C. Santiapillai, 'The Ecology of Free Living Water Buffalo (*Bubalus bubalis*) in Sri Lanka with Particular Reference to Ruhuna National Park', *Tigerpaper*, vol. 14, no. 3, 1987, pp. 20–6; R.K. Pandey, 'Habitat Utilization and Diurnal Activity Pattern of Indian Wild Buffalo (*Bubalus bubalis* L.) in Indravati National Park, India: A Study of Habitat/Animal Interaction', *Journal of Tropical Forestry*, vol. 4, no. 3, 1988, pp. 269–80; P.K. Mathur, P.K. Malik, and P.D. Muley, *Ecology and Population Genetics of Asian Wild Buffalo (Bubalus bubalis) in India*, Project Report, Dehradun: Wildlife Institute of India, 1995; P.C. Kotwal, M.L. Naik, and R.P. Mishra, 'Indian Wild Buffalo (*Bubalus bubalis* L.) an Endangered Species: Habitat and Diurnal Activity Pattern in Udanti Wildlife Sanctuary', *Cheetal*, vol. 41, no. 3–4, 2003, pp. 19–28; P.C. Kotwal and R.P. Mishra, 'Ecobiology of Indian Wild Buffalo *Bubalus arnee* L. in Udanti Wildlife Sanctuary, Chhattisgarh, India', *Journal of the Bombay Natural History Society*, vol. 101, no. 2, 2004, pp. 252–4.
9. Choudhury, *The Vanishing Herds*.
10. A.U. Choudhury, 'Wild Water Buffalo *Bubalus arnee*', in *Ecology, Evolution and Behaviour of Wild Cattle: Implications for Conservation*, ed. M. Melletti and J. Burton, Cambridge: Cambridge University Press, 2014, pp. 255–301.
11. W.R. Cockrill, 'Water buffalo', in *Evolution of Domesticated Animals*, ed. I.L. Mason, London: Longman, 1984, pp. 52–63; G.B. Corbet and J.E. Hill, *The Mammals of the Indomalayan Region: A Systematic Review*, Oxford: Oxford University Press, 1992.

12. R. Lydekker, *The Game Animals of India, Burma, Malaya and Tibet*, London: Rowland Ward, 1907; J.C. Daniel and B.R. Grubh, 'The Indian Wild Buffalo, *Bubalus bubalis* (L.) in Peninsular India: A Preliminary Survey', *Journal of the Bombay Natural History Society*, vol. 63, no. 1, 1966, pp. 32–53; H. Epstein, *Domestic Animals of China*, New York: Africana Publishing Corporation, 1971; A.K. Mukherjee, *Endangered Animals of India*, Calcutta: Zoological Survey of India, 1982; J.H. Honacki, K.E. Kinman and J.W. Koeppl, *Mammal Species of the World: A Taxonomic and Geographic Reference*, Lawrence: Allen Press Inc., and the Association of Systematic Collections, 1982; J.R. Mackinnon and K. Mackinnon, *Review of the Protected Areas System in the Indo-Malayan Realm*, Gland, Switzerland and Cambridge: IUCN, 1986.
13. Choudhury, *The Vanishing Herds*.
14. Ibid.; The IUCN Red List of Threatened Species, *IUCN* (website), 2023, see http://www.iucnredlist.org/, accessed 8 June 2023.
15. S.H. Prater, *Book of Indian Animals*, 2nd edn., Bombay: Bombay Natural History Society, 1948.
16. S. Hedges, *Asian Wild Cattle and Buffaloes: Status Report and Conservation Action Plan*, 2 vols., Working draft for CAMP/AWCSG meeting, Bangkok: WWF Project, 1995.
17. D.G. Tulloch, 'The feral swamp buffaloes of Australia's Northern Territory', in *The Husbandry and Health of the Domestic Buffalo*, ed. R.E. Cockrill, Rome: FAO, 1974, pp. 493–505; D.G. Tulloch and R.T. Litchfield, 'Wallow for Buffalo', *Australian Wild. Res*, vol. 8, 1981, pp. 555–65.
18. Zahiru'd-din Muhammad Babur, *Babur-nama* (Memoirs of Babur), tr. Annette Susannah Beveridge, vol. II, London: Luzac & Co, 1922, p. 46.
19. Ali, 'The Moghul Emperors of India as Naturalists and Sportsmen'.
20. Inverarity, 'The Indian Wild Buffalo'.
21. Choudhury, *The Vanishing Herds*.
22. Abul Fazl, *Ain-i-Akbari*, tr. Henry Blochmann (1st vol.) and Henry Sullivan Jarrett (2nd and 3rd vols), Calcutta: Asiatic Society of Bengal, 1873.
23. Maharajah of Cooch Behar, *Thirty-seven Years of Big Game Shooting in Cooch Behar, the Duars and Assam: A Rough Diary*, Bombay: The Times Press, 1908; repr. as *Big Game Shooting in Eastern And North Eastern India*, Delhi: Mittal Publications, 1985.
24. Dev P. Goswami, 'Bonoria moh rakkhar guri katha (Katha prasanga)', *Bahi*, vol. 26, no. 10, 1940, pp. 26–9.

13

The Vanishing Spirit Duck
White-winged Wood Duck

Dipanjol Deka

> The only thing we can't buy more of is time, she said, and dodo birds. We can't buy any more of them. They are extinct. And Dinosaurs.
>
> —Dean Koontz

In one of my annual ritualistic trekking and birding in Nameri National Park (Assam, India), something flew past me from a tree branch releasing an eerie wailing call. The dark densely forested area with the adjacent algae covered water body and the swampy bank made me romanticize the shadowy creature and its spooky call. As I eventually found out, it was none other than the elusive White Winged-Wood Duck (*Asarcornis scutulata*). No wonder it is known as the *deo-hanh* (spirit duck) in Assamese. I followed it in the direction it went and after a painstaking slow stalk; I managed to get a decent shot of the most elusive bird for birders worldwide. In Burma, it is called '*Mandali*', and in Indonesia '*Itik Hutan*', or Forest Duck, and in Spanish, it is known as the *Pato Almizclero Aliblanco* or *Pato de Jungla*.[1]

TAXONOMY

Kingdom	Animalia
Phylum	Chordata
Class	Aves
Order	Anseriformes
Family	Anatidae

This species' population has decreased very rapidly owing to widespread loss and degradation of its lowland riverine habitats. Its small and fragmented populations are plagued by hunting and other mostly manmade interferences which have severely disturbed its way of life. Its current population trend is decreasing, with a few scattered individual or groups worldwide. In 1951, the bird was declared to be one of the most threatened species of the ducks of north-eastern India by the Indian Wild Life Board and was placed on the special protected list. The IUCN Red List of Threatened Species, in a publication in 2017, listed the bird as Endangered (EN) per an assessment undertaken by the organization in 2016. This illusive bird was earlier also listed in the Endangered category by IUCN from 1994 to 2013, except in the year 1988 wherein it was listed as Threatened (T).[2]

IDENTIFICATION

The White-winged Wood Duck is a large duck with a black body, a white head thickly spotted with black, noticeable white patches on the wings and red or orange eyes. Its average length is 81 cm.

The male is much larger and heavier than the female and has a glossier plumage. Males have a mostly dull yellowish bill, blackish spots on the head and upper neck, white lesser and median coverts and inner edges of tertials and bluish-grey secondaries. In flight, white wing-coverts contrast with the rest of the wings and are prominently visible. Females are smaller and usually have more densely mottled head

PLATE 13.1: White-winged Wood Duck, Nameri National Park
Image courtesy: Dipanjol Deka

and upper neck. It is duller and browner in a juvenile stage. The flight call, which is best heard during the evening, is a prolonged, vibrant, wailing honk sometimes breaking into a nasal whistle at the end. The male calls with a low 'cronk cronk' and the female with a whistle. Whilst on the water, or when disturbed, a shorter harsh honk is used, often sounding as though the bird were losing its voice. When it feels threatened or it displays aggression, the birds hiss, holding the head low and, on occasion, lifting the wings to display the speculum and prominent white patches.

HABITAT AND ECOLOGY

The White-winged Wood Duck is essentially a resident of the dense tropical evergreen forest. It prefers to live in inaccessible swampy areas formed by numerous rivers, streams, crocks, etc., deep inside the forest. It is a shade-loving bird remaining in secluded jungle pools. It is one of the few ducks which often perches in trees during the day. It is a very secretive bird and often feeds only at night. If not disturbed, they may remain on the water until the sun is high, about two or three hours after dawn. Their diet consists of plant and animal material, aquatic plants, seeds of wild and cultivated plants, aquatic insects, crustaceans, molluscs, frogs, snakes and fishes.[3] The ducklings first start feeding entirely on small animals, progressively expanding their diet to include insects, worms, small snail and fish.[4]

In Nameri National Park, one has to cross the Bhorelli River early, when dawn breaks, and walk a 2.50 km. sandy stretch to the Kuruwa beel and silently creep forward towards the muddy algae-covered water body in order to glimpse the spirit duck feeding in the water. Usually found in pairs or in small groups, the species is most easily spotted at dawn when it flies back to roost or at dusk when it goes to its feeding sites. In Assam, there have been reports from fellow birdwatchers of sightings of more than ten individuals together; I personally have spotted six individuals together till now.

The White-winged Wood Duck is essentially a resident of the dense tropical evergreen forest which supports slow-flowing or stagnant water bodies. It usually inhabits inaccessible swampy areas deep inside the forest, formed by numerous rivers, streams, etc. Its optimum habitat is lowlands (below 200 m.) but it has been found up to 1,500 m. (5,000 ft.).[5] It is a shade-loving bird that prefers to live in secluded jungle pools, occasionally perching on trees during the day. Each pair of White-winged Wood Ducks needs approximately 250 acres of habitat in order to breed. It breeds in the hollows of trees during summer, at the onset of the monsoons.[6]

When roosting in the forest, they usually rely on their camouflage and try to remain undetected as they are surprisingly difficult to see in the dappled light of the forest. I have spotted and photographed the species a number of times in Nameri National Park as well as in Dehing Patkai National Park, but the birds are exceptionally shy and fly away at the slightest noise. I had, however, clad myself in camouflage clothing and painstakingly stalked it and photographed it at a close range on a couple of occasions.

GEOGRAPHIC RANGE

The species was historically widely distributed across north-eastern India and Bangladesh, through South-East Asia to Java and Sumatra, Indonesia. It has faced a dramatic decline— its population is now estimated at around 1,000 individuals, comprising 200 in Laos, Thailand, Vietnam and Cambodia, 150 in Sumatra, Indonesia, 450 in India and Bangladesh and in the 'low hundreds' in Myanmar.[7] Even though it has been recently been recorded in Bhutan, its population nonetheless continues to wane throughout its range. It is supposedly extinct in Malaysia and the latest records are from Vietnam, in Yok Don National Park, where it is rare and allegedly under recorded.[8]

In India, it has been recorded in Arunachal Pradesh, Assam, Meghalaya, Nagaland and Manipur (no recent report), with unconfirmed reports from Tripura and Mizoram. Its current distribution is chiefly in the eastern lowlands of Assam and Arunachal Pradesh.

Nameri is one of the few areas where this species has been recorded outside its main stronghold in the Dibru-Saikhowa Biosphere Reserve in Dibrugarh and Tinsukia districts, where we now estimate a population of 200 individuals (out of a total Indian population of 300–50 individuals).[9] In north-eastern Assam there are at present, known populations of White-winged Wood Duck in the reserve forests of Dibrugarh, Doom Dooma, Dehing West, Duamara and Dirok, and in the Namchick river area.[10] The confirmed population for all these areas totals only fifteen pairs, a reduction since 1969–70 of some eighteen pairs, although the Dibrugarh Reserve

PLATE 13.2: White-winged Wood Duck, Nameri National Park ©Dipanjol Deka

Forest is once again inhabited.[11] It is likely that a number of other reserve forests on the north bank of the Brahmaputra (Sadiya to Buri Suti/Jamjing area, Pobo Chopri, and east of Doom Dooma and Digboi, plus those covering plains forest in Arunachal Pradesh) also have resident populations.[12]

The habitat requirement limits the species within the great primary forest tracts of north-east India, Burma, Thailand and Indonesia.

POPULATION

There has been a lack of scrutiny of recent records, but there are an estimated 450 White-winged Wood Ducks in India, although it may be lower than this estimate. The remaining worldwide population, inclusive of Myanmar, Cambodia, Indonesia, Vietnam and Laos, may be conservatively considered to lie within the band of 250–999 mature individuals.[13]

THREATS[2]

The degradation of riverine and forest habitats is considered the main cause of decline in population of this species. Some of the decline is due to inappropriate management of forests (e.g. forest burning during the dry season). It causes loss of large trees with nesting holes which disturbs its nesting habitat. As a result, these small, fragmented populations are vulnerable to extinction from man-made interferences. In north east India, due to population pressure a large chunk of its habitat was encroached upon and destroyed. Also, due to lack of proper forest management including erosion and flood management, its dwindling habitat has suffered. One recent example is the Maniram beel in Nameri National Park, which was silted by the Bhorelli River one flood; after this, the habitat is hardly used by the duck. While trade may be more of a local threat to the species, hunting, which goes relatively unreported, is a far more serious issue.

CONSERVATION ACTIONS

In 2003, the White-Winged Wood Duck attained the status of Assam's State Bird.[14] This prompted the government of Assam to take up conservation measures with a renewed vigour.

The Wildfowl and Wetlands Trust produced an action plan for the species and is in the process of implementing it. In 1993, twenty-one protected areas were known to support populations. Dibru-Saikhowa National Park and the recently declared Dihing-Patkai National Park, both in Assam, were established because of its importance for this species, while a significant population is protected in the Nameri National Park. Some attempts were made to ensure captive breeding in the Bardobi Tea Estate (Assam) and Miao Zoo (Arunachal Pradesh), but the eggs and chicks developed tuberculosis and so no reintroduction took place.[15]

CONSERVATION ACTIONS TO BE UNDERTAKEN

Conservation activities have to be escalated if this species is to be saved from extinction. The government and other conservation organizations should conduct further surveys to clarify its distribution and status and instigate regular monitoring of selected key populations in different locations.

1. Promote strict enforcement of hunting regulations and minimize encroachment, disturbance and habitat degradation in all protected areas that support the specie's populations. The governments of all countries where this duck is found should be vigilant and, and if, necessary bring about new stringent laws to back conservation enforcement.[16]
2. Undertake campaigns and spread awareness against pesticide and oil pollution at key sites in north-east India in earnest.
3. Promote more widespread conservation awareness campaigns in and around key protected areas across nations. Rapidly introduce the measures outlined above in newly discovered strongholds, e.g. northern Myanmar.[17]
4. Ensure that captive breeding centres maintain healthy populations of this species, and ensure that diseased individuals are not able to escape and thus potentially spread diseases among wild populations.[18]

Serious efforts by conservationist organiations with strong support from the government, backed by strict enforceable laws are the need of the hour and a dire necessity sans which this wonderful bird will vanish into oblivion.[19]

NOTES

1. M.J.S. Mackenzie and Janet Kear, 'The White-winged Wood Duck', *Wildfowl*, vol. 27, 1976, pp. 5–17.
2. BirdLife International, 'Asarcornis scutulata', *The IUCN Red List of*

Threatened Species (website), 2017, see http://dx.doi.org/10.2305/IUCN.UK.2017-1.RLTS.T22680064A110103586.en, accessed 17 December 2022.

3. Niranjan Das and Sujata Deori, 'Occurrence of White-Winged Wood Duck (*Cairina Scutulata*) in Nameri National Park, Assam, India', *Bird Populations*, vol. 11, 2012, pp. 7–13.
4. Ibid.
5. Mackenzie and Kear, 'The White-winged Wood Duck'.
6. Das and Deori, 'Occurrence of White-Winged Wood Duck'.
7. BirdLife International, 'Asarcornis scutulata'.
8. Ibid.
9. Mackenzie and Kear, 'The White-winged Wood Duck'
10. Ibid.
11. Ibid.
12. Ibid.
13. BirdLife International, 'Asarcornis scutulata'.
14. Nabarun Guha, 'Heard but Rarely Seen: Plan to Save Assam's White Winged Wood Duck Takes Flight', *Indian Express*, 10 October 2019, see https://indianexpress.com/article/north-east-india/assam/the-call-of-the-spirit-bird-a-move-to-save-assams-elusive-white-winged-wood-duck-6061285/, accessed 17 December 2022.
15. BirdLife International, 'Asarcornis scutulata'.
16. Ibid.
17. Ibid.
18. Ibid.
19. Ibid.

PART III

VULNERABLE

14

The Vulnerable Beauty of the Closepet Fringes

The Yellow-throated Bulbul

M.B. Prashanth and T. Ganesh

> Birds are indicators of the environment. If they are in trouble, we know we'll soon be in trouble.
>
> —Roger Tory Peterson

The 'closepet belt' in geological terms refers to a range of hillocks that run north–south on the eastern side of the western Ghats in southern Karnataka (India).[1] The name 'Close' comes from an erstwhile British-era town named Ramanagara, which is known in ornithological circles for a small population of a critically endangered species, the Indian Vulture (*Gyps indicus*). Popularly, it is also known as the shooting site for the famous Bollywood movie *Sholay*, or for a multitude of Kannada movie locations and rock climbing activities. The Closepet granitoids are a series of hillocks that range over the southern Karnataka region and are covered with scrub, thorn and deciduous vegetation. Amidst the other renowned elements of the hillocks, such as the commercially valuable granite rock, the vultures, the temples that abound

PLATE 14.1: The boulder-strewn rocky hillock and scrub seen at the base of the hill (Madhugiri) is a favourable habitat for the Yellow-throated Bulbul. *Image source*: M.B. Prashanth

these hillocks, another important resident is the Yellow-throated Bulbul (*Pycnonotus xantholaemus*) (henceforth YTB).

DISCOVERY

T.C. Jerdon, a British-era surgeon and naturalist was the first to collect a specimen of the YTB in 1842–5, when hunters brought him a bird from the hills west of Nellore.[2] He presumed it to be extremely geographically limited. A comprehensive update of the status of the bird in the Deccan region of Karnataka, including summaries of new records from the eastern slopes of the western ghats and adjoining states, was done in the early 2000s.[3]

NATURAL HISTORY

The YTB is a small bird belonging to *Passeriformes*, an order of perching birds. Based on its habits, the bird is known to be a skulker that sticks to sparsely vegetated, thick thorn and scrub forests in the stunted rocky foothills of the Western and Eastern Ghats. A few recent records of the YTB have come from sites across the Godavari in the north, to the hills in the southern part of the Western Ghats, extending its distributional range beyond its known fringes.[4] In appearance, the YTB is closest to the White-browed Bulbul than any other bulbul species, even though its habits are known to be similar to that of the Red-vented Bulbul. The call of the bird is much mellower than the similar-looking White-browed Bulbul. The bird is known to feed on a wide range of berries and insects depending on its local habitat in the dry, thorny rocky scrub of the Deccan Plateau.[5] Some of the berries recovered from stomach analysis are that of sandal and a type of *Phyllanthus*[6] (a genus of flowering plant), and *Cipadessa bacciferra* (a species of plants).[7] The bird is chiefly known to breed in the monsoon months, i.e. from June to September.[8]

THREATENED STATUS

The YTB is classified as a Vulnerable species[9] inspite of being distributed over a large range in the Indian peninsula. Many other Threatened species are distributed over a much smaller range, such as the high hills of the Western Ghats, and may come under a protectionary purview since they are found in a national park or a sanctuary. In case of the YTB, it is the lesser-known hillocks that are quite often reserve forests or boulders amidst grazing pastures on hills. Gradual destruction of its habitat and degradation which leads to alteration of habitat is a major factor to its declining trend and Threatened status.[9]

Owing to the restricted range of the species, unlike the more widely distributed bulbul species that occur in the Indian peninsula (Red-whiskered, Red-, White-browed, Square-tailed, Yellow-browed, Black-crested, Grey-headed), the YTB has received lesser attention in terms of conservation

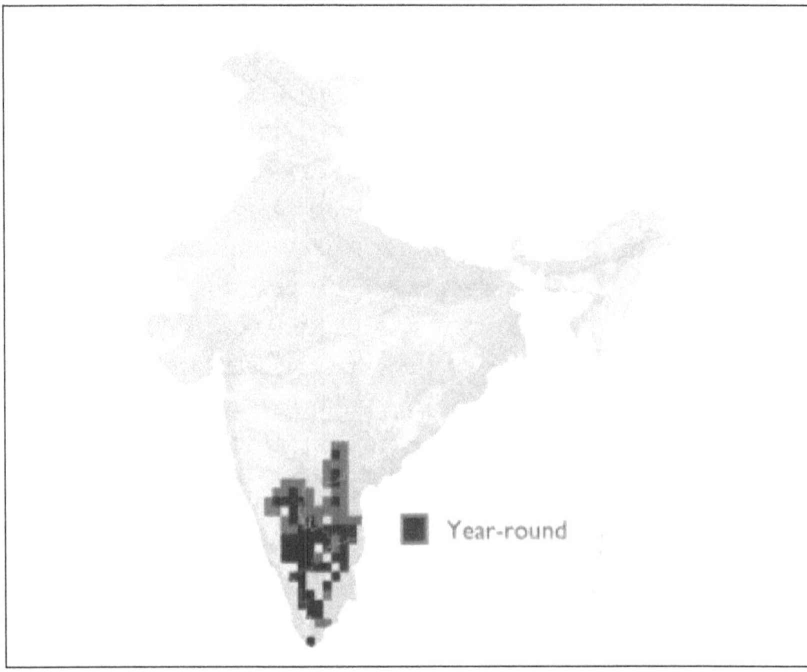

The international boundaries of India as shown in the map are neither correct nor authentic.

MAP 14.1: Distribution map of the Yellow-throated Bulbul in India
Image courtesy: SoIB 2020

measures due to the patchy nature of its habitats. The seven bulbul species mentioned above, other than the YTB, are distributed over a wide geographical area and are therefore relatively less prone to any forms of disturbance. A few of the species are extremely abundant in the country and a few others are abundant or common within protected areas. Barring a few large protected areas such as the Biligiri Ranganathswamy Temple and Wildlife Sanctuary and the Nallamalla Hills, Chinnar,[10] the YTB occurs in the smaller reserve forests and conservation reserves often recognized as Important Bird Areas, Thus it occurs in forests with a lower degree of legal protection, though it receives a lower degree of legal protection from de-notifications due to developmental pressures and clearances for mining activities.

PLATES 14.2a and 14.2b: Yellow-throated Bulbul
Image courtesy: Clement M. Francis (CC By SA 2.5)/Rudraksha Francis (CC By SA 2.5)/Wikimedia Commons

THREATS

The landscape of the hillock country where the YTB occurs is spread across a matrix of towns, villages, road networks, hillocks and scrub jungle. The roads, over the passage of time, are prone to widening and urbanization which inadvertently leads to mushrooming of urbanized clusters, leading to the gradual erosion of precious habitats. This also facilitates mass planting of unfavourable and exotic trees, annual occurrence of forest fires in summer, displacement in grazing activities, and more importantly promotion of mining granite. Mined granite from the hillocks, many of which are small monoliths with the vegetation spread between the crevices of huge rocks contours, are completely razed to the ground, which also leads to habitat fragmentation. Many such examples of completely razed granitoids occur along the major highways

PLATE 14.3: Quarrying at the fringes of boulder-strewn hills and scrub, an important habitat for the Yellow-throated Bulbul, leads to fragmentation of habitats
Image courtesy: M.B. Prashanth

leading from Bangalore towards Mysore, Tumkur, Kolar, Devanahalli and many other reserve forests. Further to the clearing of monoliths, rock contours that are widely spread along the rocky hillocks that form fringe ranges on both the eastern and western sides of the Closepet belt are prone to granite mining, stone crushers and stone ballast mining units. An instance of the YTB disappearing from an earlier recorded location due to the loss and degradation of favourable habitat has been observed at Thondebhavi.[11]

Ecological, ornithological and floral surveys that are conducted in large tiger reserves and protected areas generally skip over the YTB habitat since reserve forests are not prioritized for regular monitoring of flora and fauna but come under the purview of protection against large faunal poaching or forest land encroachments. Thus, the YTB runs the risk of becoming locally extinct in multiple smaller

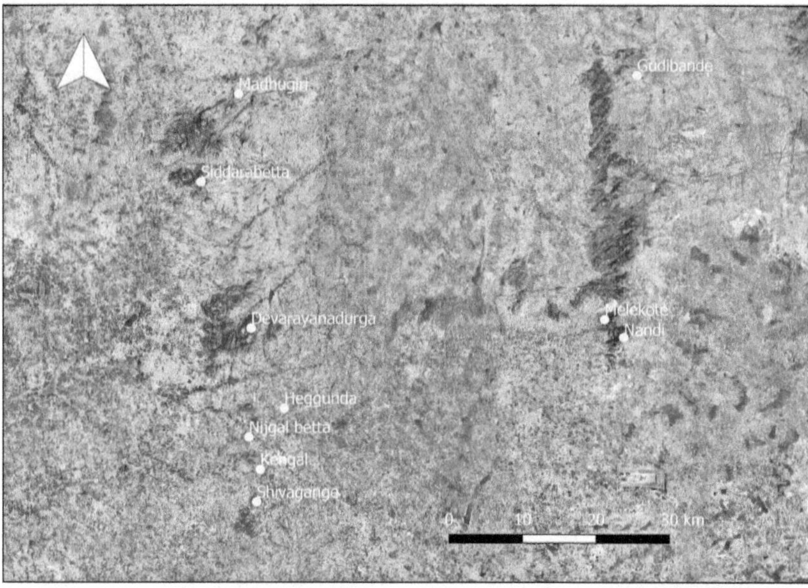

MAP 14.2: Hillocks and hill range holding habitat for the Yellow-throated Bulbul (darker shades) alongside the fragmented isolated hillocks clusters seen on the southern side
Image courtesy: Google Earth

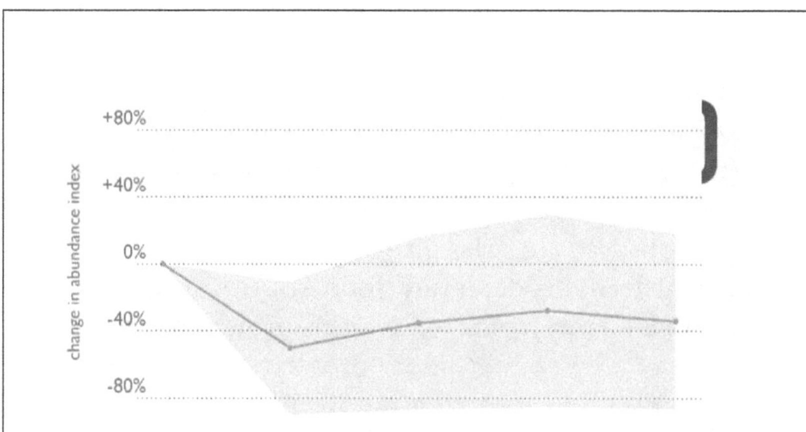

FIGURE 14.1: Trend analysis of the reported abundance of the Yellow-throated Bulbul across India from citizen science portals
Image courtesy: SoIB 2020

locations over a period of time, rather than going extinct over a large landscape. Such a process is often hard to assess or detect due to the sulking nature of the bird, wherein the bird is known to look for cover amidst the slightest hint of human movements and presence. The bird's adaptability to sites prone to a high degree of disturbance may not augur well in the event of smaller sites being totally reclaimed for mining activities. A long-term trend of the bird's abundance is available from recently compiled state of India's birds.[12] Though citizen science portals suggest a declining trend, however the boulder-strewn hillocks in remote areas of its distributional range are not frequented by birdwatchers or naturalists since they are not part of wildlife sanctuaries and national parks. Further effort in mapping the bird's range might tell us more about the declining trends for the species.

RECOMMENDATIONS FOR CONSERVATION

The steep, rocky, boulder strewn hillocks and thorny scrub country that are prone to increasing levels of habitat fragmentation and degradation will not serve well for the YTB. A notable commoner's perception of the boulder-strewn hillocks is that of a non-forested land devoid of important resourceful vegetation, large trees, or charismatic fauna such as tigers and elephants. Such notions and classifications leads to mass planting of non-native plantation species and leads to depleting food source or alteration in habitats that barely come under the protectionary lens. If such hillocks are brought under protectionary clusters from reclamation without affecting the livelihoods of dependents such as livestock grazers it will augur well for the species and prevent it from local extinction. As a suitable conservation measure, hill fragments should be further demarcated with strictly no quarrying and no mass-planting zones within the larger clusters of hillocks that form a suitable range for the vulnerable species and several other lesser-known species such as rock eagle owls, Shaheen falcons, civets, mongoose, jungle cats and others. Legislative frameworks such as conservation reserves, community reserves or even biodiversity heritage sites can be invoked for protection of the species since they often do not affect the livelihoods of locals in the demarcated reserves, thus negating conflict to a favourable extent.

NOTES

1. B.P. Radhakrishna and R. Vaidyanadhan, *Geology of Karnataka*, Bangalore: Geological Society of India, 1994.
2. Thomas Caverhill Jerdon, *The Birds of India: Being a Natural History of All the Birds Known to Inhabit Continental India, with Descriptions of the Species, Genera, Families, Tribes, and Orders, and a Brief Notice of Such Families as are Not Found in India, Making it a Manual of Ornithology Specially Adapted for India*, Calcutta: Military Orphan Press, Calcutta, 1863.
3. S. Subramanya, J.N. Prasad and S. Karthikeyan, 'Status, Habitat, Habits and Conservation of Yellow-throated Bulbul *Pycnonotus xantholaemus* (Jerdon) in South India', *J. Bombay Nat. Hist. Soc.*, vol. 103, nos. 2–3, May–December 2006.
4. R. Sreekar and C. Srinivasulu, 'New Site Record of Yellow-throated Bulbul Pycnonotus xantholaemus from Andhra Pradesh', *Indian Birds*, vol. 5, no. 5, 2010, p. 157; Praveen Jayadevan and L. Namassivayan, 'Sightings of Yellow-throated Bulbul Pycnonotus xantholaemus from Chinnar Wildlife Sanctuary, Kerala, Southern India', *Zoos' Print Journal*, vol. 21, no. 4, 2006; Chelmala Srinivasulu, 'Site records of Yellow-throated Bulbul Pycnonotus xantholaemus (Jerdon, 1844) in the Nallamala Hills, Eastern Ghats, Andhra Pradesh, India', *Zoos' Print Journal*, vol. 18, 2003, pp. 1051–2.
5. Subramanya, Prasad and Karthikeyan, 'Status, Habitat, Habits and Conservation of Yellow-throated Bulbul'.
6. S. Ali and S. Dillon Ripley, *Compact Handbook of the Birds of India and Pakistan*, 2nd edn., Oxford: Oxford University Press, 1989.
7. Personal communication with S. Subramanya.
8. M. Venkataswamappa and M.R. Chaitra, 'Observation of Nesting Yellow-throated Bulbuls', *Oriental Bird Club Bulletin*, vol. 30, 1999, pp. 32–3.
9. BirdLife International, 'Species Factsheet: *Pycnonotus xantholaemus*', *Data Zone* (website), 2020, see http://datazone.birdlife.org/species/factsheet/22712719 on 02/09/2020, accessed 17 December 2022.

10. Jayadeva and Namassivayan, 'Sightings of Yellow-throated Bulbul Pycnonotus xantholaemus'; Srinivasulu, 'Site Records of Yellow-throated Bulbul'; S. Karthikeyan, J.N. Prasad and T.S. Srinivasa, 'Yellow-throated Bulbul Pycnonotus xantholaemus (Jerdon) at Biligirirangan Hills, Karnataka', *J. Bombay Nat. Hist. Soc.*, vol. 92, 1995, pp. 123–4.
11. S. Subramanya et al., 'A Trip to Thondebhavi in Search of Yellowthroated Bulbul', *Newsletter for Birdwatchers*, vol. 30, nos. 11–12, 1990, p. 7.
12. 'State of India's Birds factsheet: Yellow-throated Bulbul, Pycnonotus xantholaemus', *State of India's Birds* (website), 2020, see https://www.stateofindiasbirds.in/species/yetbul1/, accessed 6 July 2020.

15

Greater One-horned Rhino
The Pride of Assam

Bibhab Kumar Talukdar

> The only way to save a rhinoceros is to save the environment in which it lives, because there's a mutual dependency between it and millions of other species of both animals and plants.
>
> —David Attenborough

THE GREATER ONE-horned Rhino (*Rhinoceros unicornis*), or GoHR, is currently found in India and Nepal and both countries can claim the success of securing the population of GoHRs despite challenges from illegal hunting and gradual change of grassland habitats due to invasion by other invasive plant species. The species is listed as 'Vulnerable' in the IUCN Red Lists of Threatened Animals.[1] India's Wildlife (Protection) Act 1972 listed the GoHR in Schedule 1, giving it top priority for conservation measures while CITES has listed the species in Appendix 1 since 1975, meaning trade on its body parts, including its horn, are illegal.

THE SHRINKING KINGDOM

Historically, the GoHR once existed across the entire northern part of the Indian subcontinent, along the Indus, Ganges, and Brahmaputra river basins, from Pakistan to the Indian-Burmese border, including parts of Nepal, Bangladesh and Bhutan.[2] It may also have existed in Myanmar, southern China, and Indochina, although this is uncertain. The species was common in north-western India and Pakistan until around 1600, but disappeared from this region shortly after this time.[3] The species declined sharply in the rest of the distribution range between 1600 and 1900, until it was on the brink of extinction at the beginning of the twentieth century.

The species prefers these alluvial plain grasslands, but was known to occur in adjacent swamps and forests. The populations are currently restricted to habitats surrounded

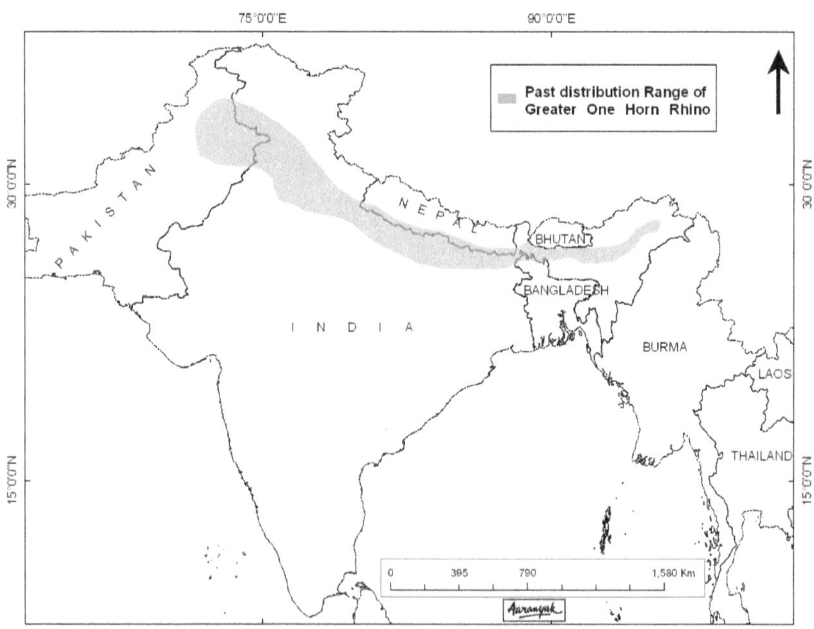

The international boundaries of India as shown in the map are neither correct nor authentic.

MAP 15.1: Historical Distribution of the GoHR. *Image courtesy:* Aaranyak

by human-dominated landscapes, so that the species is often found in adjacent cultivated areas, pastures, and secondary forests.

In India, the GoHR inhabits the riverine grasslands of the Terai and Brahmaputra basins.[4] The rhino, being a large herbivore, often creates a path through tall grassland habitats which also helps other smaller herbivores who use those rhino tracts, locally known as *dandi* (rhino path). Further, the rhino also creates wallowing sites which are also used by other smaller herbivores for their biological needs. As the grassland becomes tall and thick, smaller herbivores find it more challenging to move through those thick and tall grasslands. However, the rhino, by moving through tall and thick grassland habitats, creates a smooth path for small herbivores to move through grassland habitats. This ecological role of rhinos helps small herbivores live and share the same habitats with rhinos.

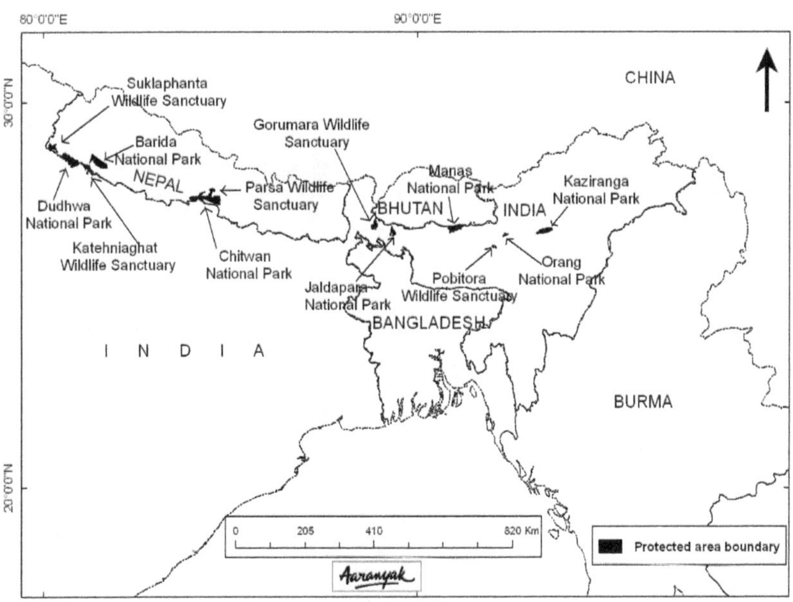

The international boundaries of India as shown in the map are neither correct nor authentic.

MAP 15.2: Present distribution of the GoHR. *Image courtesy*: Aaranyak

LIFE AND DIET

The GoHR is very fond of a variety of grasses; however, its diet also includes some fruits, leaves, shrub, tree branches, and cultivated crops.[5] The GoHR also regularly utilizes mineral licks. Males are solitary, with unstructured, overlapping territories. Females are solitary unless they have a calf. Calves remain with their mother for 3–4 years. Its life history characteristics are still little known, with longevity estimated at 35–45 years, gestation length of approximately sixteen months (as with other rhino species), and age at sexual maturity estimated at 5–7 years for females and 8–10 years for males.

ALARM IN ASSAM

With its numbers dwindling in Assam at the beginning of the previous century, it was obvious that the people of the state would develop a sense of pride in the species that symbolized Assam at the global level. And as luck would have it, the GoHR has emerged as one of the success stories in the wild; about 73 per cent of total population of the species in now found in the wild.

THE CONSERVATION STORY

For the state of Assam, the rhino symbolizes the epitome of the conservation movement and its success, keeping in view that at the beginning of the nineteenth century it was thought that about a dozen rhinos may have lived in the Kaziranga area of Assam.[6] The state government had shown its long-term vision towards conservation of forests way back in 1891 when it promulgated the Assam Forest Regulation 1891 to usher in a new era of forest conservation. The first sincere efforts to conserve the estimated dozen rhinos found in Kaziranga was initiated in the year 1905 when Kaziranga, Manas and Laokhowa were declared game reserves. Finally, in 1908 formal pronouncements were made to convert

PLATE 15.1: Greater One-horned Rhino in Kaziranga National Park
Image courtesy: Bibhab Talukdar

the three areas as game reserves. Aware of the threats from poachers, the Assam government framed the Assam Rhinoceros Preservation Act in 1954, which prohibited hunting, capture and injuring the rhinos, thus further strengthening conservation and protection of the Indian rhino in Assam. In 1968, the Assam government realized the need to convert Kaziranga Wildlife Sanctuary into a national park and made sincere effort to enact the Assam National Park Act, which came into effect since 1969. In January 1974, in pursuance of the Assam National Park Act, 1968, Kaziranga was declared the first national park in the state, and was spread over an area of about 430 sq. km. It is because of the existence of rhinos that places like Kaziranga, Manas, Orang have been declared as national parks and Pabitora, Laokhowa and Burhachapori as wildlife sanctuaries in Assam, securing the future of other wild species as well.

Since then, due to proactive conservation measures initiated in Assam, the rhino population has steadily increased to 2,885 today. As estimated in 2022, there are 2,613 rhinos in Kaziranga National Park, 125 in Orang National Park, 50 in Manas National Park and 107 in Pabitora Wildlife Sanctuary. In addition to the GoHR population found in Assam, it is worth mentioning that in West Bengal (India), there are about 339 rhinos, of which 287 are in Jaldapara National Park and 52 in Gorumara National Park in West Bengal, while in Dudhwa National Park in Uttar Pradesh 42 rhinos are estimated. In Nepal about 752 rhinos are estimated, of which 694 are in Chitwan National Park, 38 in Bardia National Park, 17 in Suklaphanta National Park and 3 in Parsa Wildlife Reserve. Today, the total wild population of GoHR is about 4018.

The state of Assam also launched Indian Rhino Vision 2020 (IRV) in 2005 with the aim of increasing the rhino population up to 3,000 in Assam by 2020 and to capture and translocate rhinos from existing rhino-bearing areas in Assam and release in other historical rhino areas.[7] Between 2008 and 2012, 22 rhinos from Pabitora Wildlife Sanctuary and Kaziranga National Park were captured and translocated to Manas NNational ParkP. No rhinos died during capture or translocation which reflects the strict scientific protocols followed by team IRV 2020. However, poaching in Manas has killed about ten rhinos since translocation started in 2008, but about twenty-two calves were born during the same period. Besides the twenty-two rhinos translocated to Manas, about eighteen rhino calves rescued during floods in Kaziranga between 2006 and 2019 were released in Manas National Park, which has also started contributing towards overall population build-up of rhinos there. To further plan future rhino conservation in Assam, a detailed scientific workshop was held in Guwahati in which Vortex Software, used for population and habitat viability analysis, was used to prepare future planning to secure GoHR populations in Assam.[8]

CONSERVATION QUANDARIES

Even with all these protective measures, rhino conservation efforts in Assam have received periodic jolts, especially during the 1980s when about seventy rhinos were exterminated by poachers in Laokhowa Wildlife Sanctuary in Nagaon district of Assam. The poachers took full advantage of the prolonged socio-political turmoil during that period. Further, Assam lost about eighty rhinos due to poaching during the 1990s in Manas National Park; this was again due to prolonged socio-political unrest in the area. These two incidences have exemplified that in times of prolonged socio-political unrest (one to five years or more), well-organized wildlife crimes take shape and target vulnerable species. The wildlife criminals take advantage of the fact that during such periods of unrest, most law enforcement agencies are hard-pressed in controlling the law and order of a particular area, which results in their attention getting diverted from other, supposedly non-alarming, matters.

MULTIPLE THREATS

When we talk about threats to rhinos, it is a popular opinion that poaching is the ultimate menace. However, although poaching is a continued threat to the rhino population, other key silent threats are: gradual spread of invasive plant species such *Ipomoea*, *Parthenium*, or *Mimosa* in grassland areas as well as weed and climber infestations by *Mikania cordata*, *Mikania scandens*, *Lantana camara* and *Leea* spp. These infestations of weeds and other invasive plant species need to be scientifically controlled so that grassland habitats are not reduced, leading to siltation in wetlands of rhino-bearing areas in Assam. This calls for periodic de-siltation to maintain a balance between wetland–grassland compositions. If food is limited in the already restricted grassland pockets in Assam, it can silently reduce the rhino population from lack of nutritious food. In such a milieu, the health of rhinos are likely to deteriorate,

PLATE 15.2: The Great One-horned Rhino in Kaziranga National Park
Image courtesy: Bibhab Talukdar

which may lead to fatality that are often enlisted as death due to natural causes! The death of each rhino needs further rigorous scientific interventions to know the actual causes of death.

THE ALMANAC MENACE: FLOODS

The annual flood in the Brahmaputra plains has regularly acted as an agent of Darwin's theory of natural, wherein weaker animals perish while the stronger ones survive to carry forward the best gene to build future populations of wild animals. In Kaziranga National Park, as the flood water rises within the core area, the animals flee to take shelter on the natural highlands in the Karbi Anglong Hills to the south of the park, often crossing the busy highway to do so. This sends the wildlife protection staff into a tizzy as they have to keep tabs on fleeing animals and regulating traffic on the busy highway becomes a challenge.

There have been discussions at the government level to construct more highlands inside the flood plain ecosystems of Kaziranga National Park so that animals need not move to Karbi-Anglong Hills or other such higher grounds outside the park's boundary during floods. However, the pertinent question is how many such highlands can be constructed within the park without disturbing its fragile natural ecosystem. In the event of soil erosion in these artificial highlands, the wetlands within the park are bound to be affected by unwarranted siltation, which may ultimately lead to shrinkage of the wet grassland spread. It is comparatively easy for rhinos to live in floodplain ecosystems, as over the decades they have survived all major floods not only in Kaziranga but also in Pabitora, with only a few casualties. This small number of casualties might not pose any threat to natural population growth of the species; however, a proper understanding of the drainage pattern of excess water of the Brahmaputra and the role of embankments along the boundary of Kaziranga and nearby areas is the call of the hour before planning any additional highlands inside the park. The best highlands for the wild animals of Kaziranga during floods is the natural Karbi-Anglong Hills. As such, conservation of the hills as part of Kaziranga–Karbi-Anglong landscape should be the obvious choice rather than inundating Kaziranga with artificial highlands. During the monsoons, strict restrictions and vigilance should be put on the movement of vehicles, especially during night time, giving wild animals ample time and space to move to Karbi-Anglong without many challenges.

CONTEMPORARY STATUS

In the state of Assam, the conservation of the GoHR has been facing renewed challenges since 2012, with 40 rhinos being poached in the year 2013. Various efforts taken by the government so far have not been able to reverse the trend of increasing rhino poaching incidents in Assam. In fact, poaching, especially in and around Kaziranga National Park, has witnessed an upsurge from September 2012, prompting the Assam government to reinforce frontline protection measures by deploying hundreds of Assam Forest Protection Force jawans with semi-automatic rifles to fight poachers. Sale of rhino horns in India is banned and at international level regulations to check illegal sale of rhino horns are being monitored by CITES.[9]

On their part, poachers follow well-planned strategies while trespassing areas where they might have found lacunae in the protection regime of the Kaziranga authorities. If one analyses the poaching trend in Kaziranga, organized poachers have targeted the eastern part of the park in Bura-pahar and Kukrakata areas. Poachers also seem to keep constant watch on stray rhinos going towards Brahmaputra Island (Chapori) outside the current area of Kaziranga. The horns of the rhino are sold in illegal international markets at exorbitant prices. As rhinos become easy prey to poachers outside the bounds of the national park, the government should strengthen protection in the fringe areas of Kaziranga where rhinos often stray. The government should form a rapid response team comprising personnel from the nearest police station and forest office to act as fast as possible to save the straying rhinos.

However, the government of Assam has already taken some landmark legal initiatives to protect rhinos in the state. The Enactment of Wildlife (Protection) (Assam Amendment) Act 2009 has facilitated substantial increase in punishment to poachers. As per this amendment, a first-time poacher will be punished with seven to ten years of imprisonment and a fine up to Rs.50,000. If the same person is convicted for the second time, he will be punished with ten years to life imprisonment and a fine of Rs.75,000. Such offences have also been made cognizable and non-bailable per this Act. Further, the government of Assam has given all forest officers the power to use firearms for protection of forests and wildlife in the state under the provision of Criminal

Procedure Code 1973. However, despite such initiatives the increasing trend of rhino poaching in Assam has not been reversed so far which is indeed a worrying issue, especially when the rhino is considered a part of our folk cultural heritage.

Regular flow of funds to rhino -earing areas must to be ensured by the government so that field officers can exercise their duties effectively. Strict anti-poaching operations need to be in place in order to reduce rhino poaching in Assam. Crime prevention is the responsibility of the government and, as such, effective and suitably intense anti-poaching operations should be supported by governments. Crime-prevention efforts should be monitored for effort, coverage and efficiency. The government must put more energy in ensuring conviction of rhino poachers in Assam to deter poaching attempts. Unless conviction is not achieved under the existing laws, reversing the increasing trend in rhino poaching will continue to remain a distant dream. For those involved in illegal killing of rhinos or trade in rhino parts, the consequences of breaking the law need to be made serious. For this successful and challenging journey of rhino conservation in Assam, besides government officials, support rendered by non-government organizations and local communities also needs special mention. Without the support of local communities and non-government organizations in the past few decades, rhino conservation would not have been possible. The, the success of rhino conservation in Assam is indeed a team effort and the same needs to continue to secure the pride of Assam.

NOTES

1. B.K. Talukdar et al, 'Rhinoceros unicornis', *IUCN Red List of Threatened Species*, 2008, pp. 1–17.
2. T.J. Foose and N. van Strien, eds., *Asian Rhinos: Status Survey and Conservation Action Plan*, Gland, Switzerland: IUCN/SSC Asian Rhino Specialist Group, 1997.

3. L.C. Rookmaker, 'The Former Distribution of the Indian Rhinoceros (*Rhinoceros unicornis*) in India and Pakistan', *Journal of the Bombay Natural History Society*, vol. 80, 1984, pp. 555–63.
4. B.P. Lahkar, B.K. Talukdar and P.K. Sarma, 'Invasive Species in Grassland Habitat: An Ecological Threat to Greater One Horned Rhino *Rhinoceros unicornis*', *Pachyderm*, vol. 49, 1022, pp. 33–9.
5. R.M. Nowak, *Walker's Mammals of theWorld*, Baltimore and London: The Johns Hopkins University Press, 1999.
6. B.K. Talukdar, 'The Current State of Rhino in Assam and Threats in the 21st Century', *Pachyderm*, vol. 29, 2000, pp. 39–47.
7. B.K. Talukdar et al., 'Re-introduction of greater one-horned rhino in Manas National Park, Assam, India under Indian Rhino Vision 2020', in *Global Re-introduction Perspectives: Further Case Studies from around the Globe*, ed. P.S. Soorae, Gland, Switzerland: IUCN/SSC Re-introduction Specialist Group, and Abu Dhabi, UAE: Environment Agency Abu Dhabi, 2013, pp 164–7.
8. S. Ellis et al., eds., *Indian Rhino Vision 2020 Population Modeling Workshop Final Report*, Final Report from the workshop held on 4–5 November 2015, Guwahati, Assam, India, Fort Worth, TX: International Rhino Foundation, 2015.
9. R. Emslie et al., *African and Asian Rhinoceroses: Status, Conservation and Trade*, A Report from the IUCN Species Survival Commission (IUCN/ SSC) African and Asian Rhino Specialist Groups and TRAFFIC to the CITES Secretariat pursuant to Resolution Conf. 9.14 (Rev. CoP15). CoP17 Doc. 68 Annex 5, Geneva: CITES Secretariat, 2016.

16

Polar Bears
The Face of Extinction

Rituraj Phukan

I don't think there's any way we can save the Polar Bears.

—Will Steger

INTO POLAR BEAR COUNTRY

My first encounter with 'Pihoqahiak', the venerated 'ever-wandering one' for the Inuit people, was in 2015 and it will remain one of my most memorable moments. About twenty minutes after the train had rolled into Churchill, I was on the van from the Churchill Northern Studies Center (CNSC). Seated in the last row, I was eager to explore the landscape of this unknown land during the 23 km. drive to the CNSC. One of the reasons that brought me to this remote town at the edge of the Canadian Arctic was the exciting possibility of polar bears. However, what ensued that day was beyond my wildest dreams.

Polar bears have always fascinated me, well before they emerged as the universal face of extinction. These powerful creatures are often used to portray the adverse impacts of global warming on the planet's wildlife. Photographs of emaciated or stranded bears are standard in media reports on Arctic warming. They are unlike any other bears, and having thrived in the harsh Arctic for millennia. Their survival is now irrevocably linked to the climate and ecological crisis.

The town of Churchill was the historic centre of the North American fur trade, where European colonialists first established the foundations for industrial exploitation of wildlife. We passed the last block of houses, with its ubiquitous polar bear warning signs, when someone screamed from the front of the van—'Polar bear!' I saw it at the same time as I heard the shout. The van screeched to a halt and I had a clear view of the animal from my window seat. The furry white giant sauntered towards the road at a steady pace. The *Nanuk*! The *Isbjorn*! The 'Old Man in the Fur Cloak!'

The bear was between the road and the railway tracks, walking in our direction. She stopped about 200 m. from the van, stood up on her hind legs, and sniffed the air. Then she turned and briskly strode away toward the station. I

PLATE 16.1: Polar bear at Churchill
Image courtesy: Rituraj Phukan

remembered the signboard outside the platform: 'Welcome to the Polar Bear Capital of the World.'

AN EXISTENTIAL CRISIS

The International Union for Conservation of Nature (IUCN) Polar Bear Specialist Group has listed the polar bear as a Vulnerable species, citing sea ice loss due to climate change as the single biggest threat to their survival. Other threats include increased commercial activities, pollution, disease, and inadequate protection of denning and seasonal resting areas.

During the two weeks I stayed at Churchill, I observed eighteen polar bears in the wild. Our research involved long hours of field work studying the impact of warming on microorganisms, nutrient contents and other wetlands dynamics, permafrost melting and biodiversity of the Tundra. In polar bear country one cannot venture out anywhere without armed guards, and for good reason. We had our work schedule disrupted by polar bears more than once.

Scientists at the CNSC attribute the spurt in polar bear intrusions in recent years to climate change. The early meltdown and late formation of ice in Hudson Bay has forced these bears to spend more time on land. With the decrease in the 'feeding window', their straying into town is common. There is a Polar Bear Alert Program to protect people, with bear warnings everywhere.

Churchill also has the only bear jail in the world—the Polar Bear Holding Facility—where intruding bears are often kept in solitary confinement. It was heartbreaking to learn that the hungry bears are fed only snow and water until conditions are favourable for their release back into the wilderness.

ARCTIC WARMING

Polar bears rely largely on seals for sustenance, and the seasonal melting of sea ice, which serves as a hunting platform for the bears, forces the bears to go ashore and fast for the warmer months. Normally, bears would fast for about four months; faster melting of sea ice means longer fasts, or the need to forage where humans live. One-third of the global population of polar bears has already felt the effects of these sea-ice changes. Studies have shown that for each week of early ice breakup, bears come ashore 10 kg. lighter and, therefore, in poorer condition.[1] Females are also giving birth to fewer cubs;[2] in fact, sightings of polar bear mothers with triplets have become rarer in the past decade.

Warmer temperatures across the Arctic cause other difficulties. More wildfires and increased rainfall both cause the collapse of polar bear maternity dens, as well as the dens of their prey—ringed seal—adding to habitat and food instabilities. Although polar bears still inhabit their historic range in nineteen Arctic subpopulations, eight of these are in decline, primarily because of global warming.[3]

After nearly four years, in 2019, I was back in polar bear country. The Svalbard archipelago has more polar bears than people and it is one of the fastest-warming regions in the world, with some areas recording an average of 5° Celsius

PLATE 16.2: Polar bear at Spitsbergen
Image courtesy: Rituraj Phukan

rise in temperatures over the past two decades. Predictably, the loss of sea and land ice has impacted the lives of the resident wildlife, including polar bears. I saw several polar bears from the expedition ship or from zodiacs while we sailed up and down the coasts of Spitsbergen.

SEA ICE LOSS

With the continuing loss of sea ice, along with additional stressors like low genetic diversity, human habitation, industrial activities, toxic substances in their food web, and reduced populations of potential prey, polar bears are likely to completely disappear from the southern Arctic ranges within thirty to forty years. Some studies predict that two-thirds of the polar bear population will be completely wiped out by the middle of this century.[4]

Although we encountered sea ice unexpectedly early, the overall extent across the Arctic had declined considerably in the past decades. Land ice had noticeable declined, and on some islands, the ice was nearly gone. According to the accompanying experts, one of the bears that we observed, was underweight and a very tense demeanour.

Sea ice is vital for the Arctic ecosystem; it is like the soil in a forest. The loss of sea ice has wide-ranging impacts on the entire ecosystem. Algae and other tiny organisms that live on sea ice are important for Arctic cod that feed on them. In turn, seals feed on the Arctic cod and polar bears prey on seals, completing the food chain.

Other changes in sea ice, such as age and salinity, also affect marine life. For example, multi-year sea ice is being replaced by first-year sea ice, which causes shifts in ice algal communities, with cascading effects on the ice-associated ecosystem.[5]

STRESSED HABITAT

The expansion of harmful toxic algal blooms, coinciding with warming oceans, threatens food sources in the Arctic region with deaths reported in birds, seals, walruses and whales.[6] The *Arctic Report Card* 2018 adds another dimension to the mounting challenges to Arctic wildlife. Microplastic contamination is on the rise, posing a threat to seabirds and marine life that can ingest debris. There has been a twenty-fold increase in marine litter in some areas of the Arctic and doubling of microplastic concentrations in others over the course of a decade. Sylvia Earle, author and marine biologist quotes:

> For humans, the Arctic is a harshly inhospitable place, but the conditions there are precisely what Polar Bears require to survive—and thrive. 'Harsh' to us is 'home' for them. Take away the ice and snow, increase the temperature by even a little, and the realm that makes their lives possible literally melts away.[7]

In February 2019, authorities in Russia's Novaya Zemlya islands were forced to declare a state of emergency after a 'mass invasion' of polar bears occurred in the settlement of Belushya Guba. Media reports and video footage show the bears eating garbage, appearing near school grounds, and entering buildings and residential homes. Such situations are increasing across the polar bear range and human–bear conflict might aggravate the survival crisis of the 'Kings of the Arctic.'

When species must travel farther than ever before to find food, they expend far more energy than necessary. They also must compete with, or otherwise interact with, new species they have never encountered before, which can seriously disrupt the Arctic's delicate marine ecosystems.[8]

In 2015, a week before my arrival, a grizzly bear was seen for the first time ever at the CNSC. A week before that, there was another another rare grizzly sighting at Wapusk National Park. There was no evidence of grizzly bears in the area before 1996, not even in the trapping data from centuries of Hudson Bay Company operations; however recent records published in Canadian Field Naturalist show increasing frequency of sightings.[9] It is now established that grizzlies have extended their range northward, bringing them in contact with polar bears, who are being forced to spend more time on land as well due to loss of sea ice.

When grizzlies and polar bears meet, they compete for food, but they might also mate. The hybrids are known by various portmanteau names (pizzly, grolar bear, polizzly, etc.). In Canada, *nanulak* is preferred, from the Inuit *nanuk* for polar bear and *aklak*, for grizzly. Unfortunately, hybridization has emerged as another threat to the polar bear population according to University of Alberta bear biologist Andrew Derocher, who believes the polar bear population may lose out genetically.[10]

REGIME SHIFTS

The decline in polar bear population will also affect the Arctic fox, as their existence is mutually beneficial. The Arctic fox is regarded as a prime indicator of the overall health of the Tundra, due to its sensitivity to the habitat and the fact that the animal is unlikely to survive in any other biome. The small creature is highly adapted to its habitat, with thick fur, specialized heat-retaining circulatory system in its feet, and the ability to lower its metabolic rate—factors that would be detrimental in warmer environments.

To estimate the effect of climate change on a species, scientists use what is known as a *climatic or bioclimatic envelope*—a range of temperatures, rainfall, and other climate-related parameters, in which a species currently exists.[11] Wildlife species tend to move toward the edges of their climatic envelopes if climatic conditions are not right. The polar bear, for example, will migrate to cooler, moister environments, usually uphill or toward the North Pole. In many cases, however, such migration proves impossible because of unfavorable environmental parameters, geographical or human-made barriers, and competition from species that already inhabit the new area.[12]

The Arctic Council has documented ecosystem changes across the entire region and noted that these changes are accelerating in response to physical climate changes.[13] The council's biodiversity working group, the Conservation of Arctic Flora and Fauna,[14] reports that there has been disappearance or dramatic modification of habitats, ecosystems, and populations; shifts in their geographic ranges; changes in the timing of ecological events like flowering, migration, and breeding; and outbreaks of pests and disease affecting plants and animals.

The Council's 2016 *Arctic Resilience Report*[15] describes nineteen regime shifts, defined as large and persistent changes in the structure and function of social-ecological systems, which have already made an enormous impact on native wildlife, climate stability, and the people who depend on both. It is a foregone conclusion that the polar bear faces a grim future in the warming Arctic.

PLATE 16.3: Polar bear at Churchill
Image courtesy: Rituraj Phukan

Meanwhile, in a recent study published in June 2022,[16] researchers have revealed existence for the world's twentieth sub-population of polar bears of around 300[17] individuals. This evidence shows a genetically divergent species, one that has been 'functionally isolated' in south-east Greenland for almost 200 years.[18] This population seems comparatively less dependent on sea ice, and more on glacial mélange on the Alaskan fjords, throwing light on the adaptive capacity of the species at a time when the entire Arctic habitat is changing due to an alarming loss of sea ice.

ARCTIC WARMING AFFECTS ALL

Changes in the Arctic environment have sweeping consequences for the rest of the world and may also foreshadow what could happen elsewhere on the planet. Understanding how those changes, including warming temperatures and melting sea ice, have generated marine and terrestrial biodiversity losses in the Arctic is only the first step. Looking at those losses on a planet-wide basis, and their importance, is critical to our collective responses and reactions to climate change in general.[19]

The extinction of any keynote species will have catastrophic cascading impacts on the ecosystem. In the case of the Arctic, it is even more pertinent. It is believed that 'what happens in the arctic doesn't stay in the arctic, it affects us all'. In ensuring that the iconic polar bear is saved from oblivion, mankind will have scripted the roadmap for its own survival on earth.

NOTES

1. M.P. Galicia et al., 'Dietary Habits of Polar Bears in Foxe Basin, Canada: Possible Evidence of a Trophic Regime Shift Mediated by a New Top Predator', *Ecol. Evol.*, vol. 6, 2016, pp. 6005–18.
2. Louise Gray, 'Polar Bears are Having Fewer Cubs Due to Global Warming', *Telegraph*, 8 February 2011, see http://www.telegraph.

co.uk/news/earth/environment/climatechange/8311137/Polar-bears-having-fewer-cubs-due-to-global-warming.html, accessed 19 December 2022.

3. Rituraj Phukan, 'Biodiversity in a Warming World', in *Climate Abandoned: We're on the Endangered Species List*, ed. J. Cody, Carmel, CA: Writing Endeavors Press, 2019, pp. 11–24.
4. N. Lunn et al., 'Demography of an Apex Predator at the Edge of its Range: Impacts of Changing Sea Ice on Polar Bears in Hudson Bay', *Ecol. Appl.*, vol. 26, 2016, pp. 1302–20.
5. *State of the Arctic Marine Biodiversity: Key Findings and Advice for Monitoring*, Akureyri, Iceland: Conservation of Arctic Flora and Fauna International Secretariat, 2017.
6. D.M. Anderson, M.L. Richlen and K.A. Lefebvre, 'Harmful Algal Blooms in the Arctic', *Arctic Report Card*, 2018, see https://www.arctic.noaa.gov/Report-Card, accessed 19 December 2022.
7. See https://www.brainyquote.com/quotes/sylvia_earle_540315.
8. Carlos M. Duarte, *Impacts of Global Warming on Polar Ecosystems*, Bilbao, Spain: Fundación BBVA, 2008.
9. Lunn, 'Demography of an Apex Predator'.
10. T.C. Atwood et al., 'Forecasting the Relative Influence of Environmental and Anthropogenic Stressors on Polar Bears', *Ecosphere*, vol. 7, 2016.
11. Phukan, 'Biodiversity in a Warming World'.
12. Céline Bellard et al., 'Impacts of Climate Change on the Future of Biodiversity', *Ecology Letters*, vol. 15, no. 4, 2012.
13. M. Carson and G. Peterson, eds., *Arctic Resilience Report*, Stockholm: Stockholm Environment Institute and Stockholm Resilience Centre, 2016.
14. *State of the Arctic Marine Biodiversity*.
15. Carson and Peterson, *Arctic Resilience Report*.
16. Elizabeth Peacock, 'A New Polar Bear Population', *Science*, vol. 376, 2022, no. 6599, 2022, pp. 1267–8.
17. Harry Baker, 'A Secret Polar Bear Population Has Been Found in an "Impossible" Location', *Science Alert* (website), 2022, see https://www.sciencealert.com/polar-bears-discovered-secretly-living-in-seemingly-inhospitable-habitat, accessed 19 December 2022.
18. Ibid.
19. Phukan, 'Biodiversity in a Warming World'.

PART IV

NEAR THREATENED

17

The Threatened Emperor

Kaiser-i-Hind

Monsoon Jyoti Gogoi

> If all the insects were to disappear from the earth, within 50 years all life on earth would end. If all human beings disappeared from the earth, within 50 years all forms of life would flourish.
>
> —JONAS SALK

INSECT DEPLETION ACROSS THE WORLD

IT IS ESTIMATED that more than 80 per cent of all animal species on earth are insects.[1] Many insect species are under threat from anthropogenic drivers of global change. In recent decades, entomologists (someone who studies insects) have documented alarming declines in occurrence, taxonomic richness, and geographic range of insects around the world. However, important aspects of the insect decline phenomenon remain largely unknown. It is difficult to know what we are losing when 80 per cent of insect species (representing, conservatively, four million species) remain undescribed and their natural histories unknown. Most extinctions estimated to have occurred in the historical past, or predicted to occur in the future, are of insects. Despite this, the study of insect extinctions has been neglected. Only 70 modern insect extinctions have been documented, although thousands are estimated to have occurred.[2]

LOCAL EXTINCTION OF BUTTERFLIES

Butterflies are day-flying insects belonging to *Lepidoptera* and are considered flagship taxa for invertebrate conservation. There are over 18,000 species of butterflies in the world and are the best-recognized insects due to their diurnal habitats and their bright colours, spectacular shapes and smooth flight. The Indian sub-region hosts about 1,504 species of out of which, the North-East has 962 species of butterflies.[3] The Himalayas are home to around a thousand species of butterflies and many of them are endemic with narrow distribution ranges. Their exact numbers are still unknown. Roy's Argus (*Callerebia dibangensis*) is such butterfly species endemic to Dibang valley, and was discovered only in 2013. In the eastern Himalaya and north-eastern India, an example of local extinction of butterflies is Kaiser-i-Hind (*Teinopalpus imperialis*). The Kaiser-i-Hind is a much sought-after butterfly for its beauty and rarity. It is protected by Indian law under Schedule II of the Wildlife (Protection) Act, 1972. Among the rare and threatened butterflies of world, species such as the Homerus Swallowtail, Queen Alexandra's Birdwing, the Kaiser-I-Hind, and the Apollo are so spectacular that humankind surely has a major responsibility to ensure their survival and recovery to safe and productive population levels.

The Kaiser-i-Hind is one of the most elusive swallowtail butterflies, with about a 90–120 mm. wingspan. The common name literally means 'Emperor of India'. The spectacular swallowtail butterfly range is restricted to high elevation and intact forests and the present day, such forests are a rarity. The Kaiser-i-Hind is a superb green, black and orange-tailed butterfly. The green iridescence of the wings are due to the three-dimensional photonic structure of the scales and is the subject of much research. Although widely distributed, it is very local and rare.

The Kaiser-i-Hind was confined only to a small area of the Khasi Hills (Meghalaya), and due to its rampant export business, it was on the road to extinction.[4] As such, today the species is extinct in Meghalaya.[5] The species is only confined to very few pockets of Arunachal Pradesh, Manipur and West Bengal (Neora valley) and if not conserved, the species will also go extinct from these states in India.

DISTRIBUTION AND HABITAT

The Kaiser-i-Hind is found in small pockets of Nepal, Bhutan and along the eastern Himalayas in India (West Bengal, Meghalaya, Assam, Sikkim and Manipur). It is also found in northern Myanmar, northern Vietnam, north Thailand and in the Sichuan province of China. It is mostly seen above 6,000 ft. (1,800 m.). It is found at high elevations between 1828–3048 m. in wooded areas. Its flight is strong and rapid and it keeps to the top of trees; therefore, it is difficult to capture. It was once plentiful in the Khasi Hills, near Cherrapunjee but now extinct in Meghalaya.

In Nepal, the species is found in high altitude regions up to southern Burma and in the Chinese provinces of Sichuan (as *T. behludinii*) and Hubei. Records of the species are also found in Yunnan. The records from Guangdong refer to it as *T. aureus*. The subspecies *T.i. imperialis* occurs between 2000 and 3500 m. in Nepal, across north-eastern India and Bhutan to northern Burma and at lower altitudes (over 1000 m.) in China. In India, it has been recorded from Sikkim (only one dead specimen in the last twenty years), the Darjeeling region of northern West Bengal, Arunachal Pradesh, north of Cherrapunji in Meghalaya (now extinct) and Manipur in the North-East. It is, however, absent in Nagaland. Another subspecies, *T.i. imperatrix*, occurs only in Myanmar (Shan states to upper Tenasserim), north Thailand, Laos, and Vietnam, where its distribution records indicate an extension to lower altitudes than the nominate subspecies. Although there have been no published records of *T. imperialis* in the highlands of western Thailand, its presence there is a possibility. In China, *T. imperialisis* is possibly fairly widespread, perhaps occurring in the montane plateau of Yunnan and Sichuan, eastward to the highland parts of Hubei. Unfortunately, there have been no reports from China for almost a decade.[6] The species varies considerable in wingspan, *ssp. imperatrix* being much larger than the typical subspecies. Furthermore, in *ssp. imperatrix*, the yellow spot on the hindwing is usually bigger, especially in females.

ECOLOGY OF THE KAISER-I-HIND

The Kaiser-i-Hind is restricted to mountainous wooded districts, where it flies high into the canopy of trees.[7] It is said to descend nearer the ground from 8.00 to 11.00 a.m. on sunny days, when it can be attracted with bait.[8] Its flight is very rapid, particularly the male, and the best places to observe it are open places and mountaintops surrounded by forests.[9] The butterflies fly up through the forest to the hilltops in search of a mate. Often, the males will defend mating territories on the summits, fighting off other suitors. In cloudy or misty weather, the butterflies rest on low bushes, concealed by the ventral camouflage of their wings. Adults have not been observed to visit either flowers or water. The caterpillar is green with a large head and the pupa is smooth, green and horned. Larvae pupate at the end of September and adults emerge in April–May in Sikkim, and May–July in Manipur. There are reports that the species is double brooded, at least in India. Their diet consists of *Daphne spp.*, small laurel-like trees or shrubs in the *Thymeleaceae*. The habitat in which *T. imperialis* is found consists mainly of broad-leaved evergreen forests, sometimes mixed with conifers and often well-stocked with various species of oak (*Quercus*). The foodplants on which they feed are generally a component of the under story.[10]

TAXONOMIC STATUS

Kaiser-i-Hind is a large, beautiful tailed butterfly with sexes that differ in size and coloration. Because of this difference, they were originally described as two separate species. There are two (or perhaps three) subspecies. The wingspan of the nominate subspecies is 90–120 mm. (male forewing length 52 mm., female 60–5 mm.); the female *T.i. imperatrix de Niceville* is even larger, with a wingspan of 130 mm., a forewing length of 70 mm, and five distinct tails. *T. behludinii* from Sichuan is almost certainly referable to *T. imperialis*, but the type specimen is apparently untraceable and no other specimens are known. New material is needed to establish *behludinii* as a good species. Until this is forthcoming it may be assumed that *behludinii* is a subspecies or population of *imperialis*. The underside forewing (UFW) on males is black, densely suffused with bands of green scales. The underside hindwing (UHW) is scalloped, with a long golden-tipped tail. The green bands of the UFW continue, but the distal region has a bright orange patch and a series of submarginal lunules in orange and bright green. The lower forewing (LFW) basal area is densely covered with green scales, and the distal is two-thirds yellow-brown with narrow black bands. The lower hindwing (LHW) is similar to the UHW but with broader yellow markings. Females resemble the male in the UHW basal area but the outer two-thirds of the wing has broad, indistinct alternate grey and green bands. The UHW is deeply scalloped with two long, black tails and a shorter one between them. The yellow discal patch of the male is replaced by a much larger dark grey patch with yellow markings beneath. The LFW/LHW is similar to the male but with the yellow-brown colour replaced with grey and with a broad ochreous submarginal band on the LFW.[11]

SIGHTING IN THE LAST TWENTY YEARS

In 2012, the first dead specimen of Kaiser-i-Hind was recorded in Sikkim by Usha Lachugpa, a senior forest official.

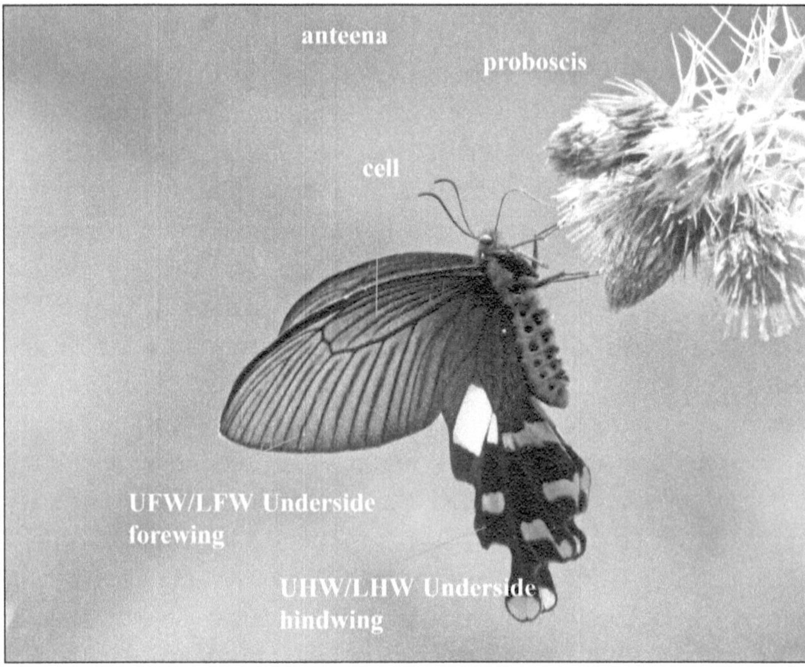

PLATE 17.1: Image showing details of the butterfly's wings
Image courtesy: Monsoon J. Gogoi

In 2013, a worn-out specimen of the species was found in Manipur by Baleshwor Soibam.[12] The species was commonly found sixty years ago in Meghalaya, but probably extirpated due to habitat loss and the export business of the specimen as trophies. Kaiser-i-Hind is probably the most elusive swallowtail of the eastern Himalaya and north-east India due to the fact that they largely remain among the treetops and camouflage themselves among leaves. Their habitat is, therefore, wooded hill forests at a high elevation, which are, unfortunately, ever declining. The species shows the importance of high-elevation wooded forests; hence, such habitats must be conserved. Talle valley of Ziro, Eaglenest Wildlife Sanctuary of West Kameng, Koubru Mountains of Manipur and Neora valley of Sikkim are few remnants of such type of forest. The butterfly also increases the potential of butterfly-based community tourism in Arunachal Pradesh.

DEMAND IN THE INTERNATIONAL MARKET IN THE 1980s

From the 1950s to the 1980s, high prices were paid for species that were rare and difficult to obtain. Species from inaccessible countries rarely appear on the market and when they do they tend to have a high price. During the 1980s, the Kaiser-i-Hind was in great demand in butterfly parks and museums across the world; as such, a large number of these were smuggled out of the country, pushing it to the verge of extinction. Around the foothills of the Himalayas, the Kaiser-i-Hind is difficult to capture since it is a strong flyer that keeps to the treetops. In 1955, a collector paid the equivalent of about $250 for a single specimen, which now costs up to $75 a pair. There are reports from Phulchoki Mountain that encampments of foreign entomologists wait for the Kaiser-i-Hind on hill tops and kill every specimen that flies in. The species had garnered attention due to its high price in the international market, because of which it is threatened with extinction,[13] Even though the Kaiser-i-Hind is protected by Indian and Nepalese law, enforcement of these protections in these countries is not effective, resulting in its continued hunt. The two species of Kaiser-i-Hind were listed in CITES Appendix II from 1987 onwards.

HABITAT UNDER THREAT

The species' habitat across its range countries is under severe pressure from logging and human activities. Four main factors threaten the Kaiser-i-Hind: habitat destruction, pollution (which might be considered a special form of habitat destruction), commercial exploitation and climate change. An effort should be made towards conserving the species and its habitat to re-establish its population.[14] The tradition of deforestation and of setting wildfires in the mountains ranges and hills should be stopped by holding awareness campaigns. Encouraging the planting of trees like Campbell's Magnolia campbellii (*Magnoliaceae*) and Nepalese

Paper Plant Daphne bholua (*Thymelaeaceae*),[15] the known host plants for the larvae of the Kaiser-i-Hind, will help the species' population and habitat to grow.

In general, the threats to Kaiser-i-Hind are either the direct result of increasing human population pressure or else are enhanced by it. It is now internationally recognized that conservation efforts can only reach long-term fruition in partnership with policies for population planning and control. Some of the forests still retain their original character, but human population pressure is increasing their degradation through fire, deforestation, grazing and fuelwood collecting. The global problem of human populations outstripping their resources, and of the underlying political, economic and social causes, is beyond control. The Kaiser-i-Hind is a keystone species for conservation of high-altitude forest. Conservation in India is complicated by the fact that most of the responsibility for conservation action falls on the state and not on the central government. There is a need for collecting more data in the habitat area and surrounding areas such as the adjoining mountain regions of China where the species is likely to occur.

For its survival, the species appears to be dependent upon the natural forests that harbour its food plant Daphne in the Himalayas, Burma and China. There is no evidence of its spread into degraded forests, fields that are in the regenerative stages of shifting cultivation, or agricultural land. Nothing has been published concerning threats specific to the butterfly, but there is growing literature describing the disturbing extent of deforestation over the entire area of the butterfly's range. In Nepal, deforestation has largely resulted from rising human population and migration into the Terai or southern lowland strip, where broad-leaved forests grow. Loss of these lowland forests is running at an estimated 80,000 ha. per year, almost 5 per cent of the total forest cover in 1980. The migrants come from hill areas, which are more difficult to cultivate. Virtually all of central Nepal is now deforested, although certain spots are known to contain *Teinopalpus*, such as Phulchoki Mountain. In India, the Himalayan forests are possibly being deforested more slowly than in Nepal, although the available data is known to be unreliable. The isolated montane area of Meghalaya seems to be particularly seriously deforested. Koubru Mountain in Manipur is highly disturbed through anthropogenic disturbances like logging, agriculture, and wildfire for the cultivation of crops. Such activities are the major threat to the existence of the species in the area. The traditional form of agriculture her was shifting cultivation, or 'jhumming', but over-exploitation has led to severe soil erosion.[16] Such degradation in the lower-altitude regions has undoubtedly led to more severe environmental pressures and forest degradation in the high-altitude core of Meghalaya. In Darjeeling, the forests are reported to be virtually all gone and in Sikkim also large areas are denuded and eroded. In Bhutan, the forests are said to be subjected to heavy biotic interference, particularly grazing and fires. In Burma, deforestation is relatively limited, but again the main pressure on forests is the escalating human population and the resultant increased need for fuelwood, grazing and land for shifting cultivation.

Insufficient data is available for the Chinese part of the butterfly's range; specimens have been recorded as low as 1000–1400 m., where habitat destruction is likely to have been severe. Information on the degradation of the Himalayan forests is scarce and often out of date. It is thus difficult to obtain an accurate impression of human impact on the butterfly's habitat, but the indications are that its range must be shrinking rapidly.

Along with supporting important flora and fauna, not to mention human communities, these forests have important climatic and environmental influences. Most of the important rivers of northern India originate in the Himalayas. The river Brahmaputra has its headwaters in the habitat of *T. imperialis*. Biotic interference is often serious in the lower zones of hardwood forest, declining higher up. Fires may be devastating at any altitude, and grazing and fuelwood collection are on the rise. In general, most of the high-altitude coniferous forests seem to have retained their original character, particularly in the Himalayas, but increasing population pressure may eventually lead to intensified use of the land. Faced with this eventuality, great skills will be needed if environmental degradation is to be avoided. Despite the wide range of *T. imperialis*, some of the reports of over-collecting are a matter for concern. For example, in July 1984 on Phulchoki Mountain near Kathmandu in Nepal, every specimen flying in was being caught by foreign collectors. Although such an isolated event is unlikely to have a serious impact on the butterfly species, other hills may be witnessing similar conduct, even as collecting is illegal without a licence[17] in both Nepal and India.

CONSERVATION MEASURES

Specific measures for the Kaiser-i-Hind are difficult to propose at this stage. The legislation already enacted in India and Nepal should be enforced more effectively. The best way to preserve the Kaiser-i-Hind is rational utilization and protection of the forests in which it lives. The government of Nepal is preparing a National Conservation Strategy that will be of benefit to all wildlife. In India, such strategies would have to be on a state-wide basis, since each state is virtually autonomous in its forestry practices. The presence of *T. imperialis* in protected areas should be noted, and research towards its management measures encouraged, as done in the Talle valley of Arunachal Pradesh. The proposal to declare Kaiser-i-Hind the state butterfly was accepted by the State Wildlife Board of Arunachal Pradesh in 2020. The species is only confined to very few pockets of Arunachal Pradesh and, if not conserved, is sure to go extinct. The decision to declare it the state butterfly is itself a step to save the species from extinction Hopefully, the Emperor will not remain 'Threatened' in the near future.

NOTES

1. N.E. Stork, 'How Many Species of Insects and Other Terrestrial Arthropods Are There on Earth?', *Annu. Rev. Entomol*, vol. 63, 2018, pp. 31–45.
2. R.R. Dunn, 'Modern Insect Extinctions, the Neglected Majority', *Conservation Biology*, vol. 19, no. 4, 2005, pp. 1030–6.
3. W.H. Evans, *The Identification of Indian Butterflies*, 2nd edn., Bombay: Bombay Natural History Society, 1932.
4. R.E. Parsons and K. Cantlie, 'The Butterflies of the Khasia and Jaintia Hills, Assam', *Journal of the Bombay Natural History Society*, vol. 47, 1948, pp. 498–522.
5. C.B. Antram, *Butterflies of India*, Calcutta and Simla: Thacker, Spink & Co., 1924.
6. N.M. Collins and M.G. Morris, *Threatened Swallowtail Butterflies of the World*, The IUCN Red Data Book, Gland and Cambridge: IUCN, 1985.
7. G. Talbot, *The Fauna of British India, including Ceylon and Burma*, vol 1: Butterflies, London: Taylor and Francis, London, 1939.
8. Ibid.
9. K. Jordan, 'Papilionidae, Papilio to Armandia', *Macrolepidoptera of the World*, vol. 9, 1908–9, pp. 11–109.
10. Collins and Morris, *Threatened Swallowtail Butterflies of the World*.
11. Evans, *Identification of Indian Butterflies*; Antram, *Butterflies of India*; Talbot, *Fauna of British India*.
12. B. Soibam, 'Recent sightings of Kaiser-I-Hind Teinopalpus imperialis Hope, 1843 (Lepidoptera: Teinopalpani) from Manipur, India', *Journal of Threatened Taxa*, vol. 8, no. 6, 2016, pp. 8930–33.
13. Collins and Morris, *Threatened Swallowtail Butterflies of the World*.
14. T. New, R. Collins and N. Mark, Swallowtail Butterflies: An Action Plan for Their Conservation, IUCN/SSC Species Action Plans Series, Switzerland: IUCN SSC Lepidoptera Specialist Group, 1991, p. 36.
15. I. Kehimkar, *The Book of Indian Butterflies*, Bombay: Bombay Natural History Society and Delhi: Oxford University, 2008; B. Khanal, M.K. Chalise and G.S. Solanki, 'Threatened Butterflies of Central Nepal', *Journal of Threatened Taxa*, vol. 5, no. 11, 2013, pp. 4612–15.
16. Soibam, 'Recent sightings of Kaiser-I-Hind'.
17. Collins and Morris, *Threatened Swallowtail Butterflies of the World*.

Conclusion

Success Stories on the Frontline of Conservation

> We don't inherit the earth from our ancestors; we borrow it from our children.
>
> —NATIVE AMERICAN PROVERB

> If the bee disappeared off the face of the Earth, man would only have four years left to live.
>
> —ALBERT EINSTEIN

BETWEEN A FIFTH and third of all species on this planet will be extinct by 2050, while 'we' have already caused the disappearance of 1 out of every 25 lizards due to the ever-increasing heat.[1] The official addition of over 20,000 species of bees in the list of Endangered species, with a massive 89 per cent drop in numbers between 2007 and 2016, is also alarming.[2] In june 2019, one million species of plants and animals were at risk of extinction. At least 571 species have been lost since 1750.[3] The main cause of these extinctions is the destruction of natural habitats by human activities, such as cutting down forests and converting land into fields for farming.[4] According to an alarming report presented in 2019 by the Intergovernmental Science-Policy Platform on Biodiversity and Ecosystem Services (IPBES),[5] in parts of the ocean ecosystem there is very little life. Many far-flung tropical forests are practically silent as insects have vanished, and grasslands are steadily turning into deserts. Anthropocentric activity has resulted in glaring changes in more than 75 per sent of the Earth's land areas, per global assessment. Around 66 per cent of the oceans, covering most of our blue planet, have suffered substantially due to human activities. This comprises more than 400 'dead zones'—where scant or no life can survive—that collectively would be bigger than the entire United Kingdom. This new report gives us 'an ominous picture' of the health of our ecosystems and acts as a clarion call for humanity, which is the cause of its own impending doom.

So, why is this happening worldwide? We are 'precipitating an extinction spasm of a magnitude not seen since the last mass extinction event 65 million years ago', as we have been 'growing at around 1 billion people every 12 years. That's [the] equivalent to a new Athens- or Nairobi-sized city every Month'.[6] Our unquenchable demand for higher standards of living, intrinsic human selfishness and our growing disconnect with nature are the broader foundations on which all such injustices against nature seem to be based.

The earth has journeyed millions of years to witness the yielding of life that now dwells on it—it has been an unbelievably extensive journey of multifarious levels of evolving and diversifying biomes reaching a state of adjustment and equilibrium with its environments. The environs, on the other hand, kept meticulously determining and guiding the life it sustained, while enclosing paradoxical elements that were hostile as well as auxiliary. Given time—time not just in years but in epochs—life adjusts, and a balance has been attained. However, in just a couple of centuries, more so, in the past few decades, anthropocentric changes in the biosphere have almost surpassed the past historic evolutions. And human, sadly, as always, fails to acknowledge this and the fast approaching self-doom that is just around the corner.

In June 2020, Assam witnessed the Dehing-Patkai and Maguri debacles, calamities meted out on nature because of human greed. This goes back to the decision of the Central Government of India to legalize open-cast coal mining in Saleki—a proposed reserved forest—which comes under the Dehing Patkai Elephant Reserve. Found in the Indo-Burma biodiversity hotspot, Dehing Patkai is deemed to be the Amazon of the East.

Saleki has a delicate ecosystem. It is a protected elephant reserve but also houses a range of wild flora and fauna. This includes the Vulnerable king cobra and the Threatened Asian leaf turtle. The White-winged Wood Duck (the state bird of Assam), the Fox-tail Orchid (the state flower) and the Hollong (the state tree) can only be found in the Dehing Patkai Reserve.

The Maguri *beel*, on the other hand, is a vast wetland in Tinsukia (Assam) and home to several rare bird species. It also acts as a sojourn for many species from around the globe; some of the migratory species being the Ruddy Shelduck, Bar-Headed Goose, Falcated Duck, Baikal Teal, Glossy Ibis, Northern Pintail, Ferrigunuous Duck, Eurasian Wegion, Common Teal, Eurasian Curlew, Black- Headed Ibis, etc.[7] It is, 3.80 km. away from Guijan Ghat, which is, incidentally, the doorway to Dibru-Saikhowa National Park and Biosphere Reserve as well. It serves as a major attraction for ornithologists and bird lovers from all over the world, more so after BirdLife International declared it a vital Birding Site (IBA).[8] As a comprehensive ecosystem, the *beel* is bountiful in aquatic life as well as a variety of grassland birds due to the grasslands surrounding it.

Sadly, the grassland has been lately inundated by oil rigs, leading to serious disruption of ecological balance in the area.[9] Making matters worse, on 27 May 2020 there have been two major blasts in two oil wells.[10] This has led to a series of disastrous repercussions,[11] with thousands of families, their crops and livestock being devastated, heavy loss of grassland flora and fauna, and endless minor earthquakes. The grassland area of the wetlands was completely burnt up to a radius of 1–2 km. Due to these fires, Maguri burnt for over a month.

Since 2019, deforestation and fires have led to the Amazon losing approximately 10,000 sq. km. of forest cover every year.[12] While experts, until very recently, accounted for only the vegetation loss in the affected areas, the biodiversity loss caused by such fires has never been assessed. However, a 2021 study[13] states that 93–5 per cent of the 14,000 species of animals and plants have been affected, in some way or the other, by the fires in the Amazon.

The world is in need of leaders to demonstrate that the present can become a pleasant history. Stories of dedicated angels who have made a space, physical and cerebral, for beings other than humans are like the silver lining on a dark cloud. The conviction with which they pursue their quest is praiseworthy and exemplary, for it is this intent that will save the world; the result is simply an afterthought.

While most books on wildlife conservation deliberate on stories of gloom and doom, this book, along with discussing the same, aims at ending on a positive note. Let us contemplate if nature's continued loss is inescapable, or if there are grounds for hope. As much as awareness of our losses is important, perhaps conservation could benefit from dwelling more on its successes—from rejoicing in their triumph, learning from them, and using this to inspire and empower people to do more.

Whereas each conservation story has its own quintessential characteristics, all of them offer some consistent insights into how projects can have greater successes than failures, or how ordinary people can bring about extraordinary change.

THE HARGILA FIGHTERS

The Greater Adjutant Stork (leptoptilos dubius) or Hargila (bone swallower), as it is locally known in Assam (India), is listed as 'Endangered' in IUCN Red list of threatened species. It is currently known to breed only in Assam and Cambodia.

In the beginning of the twentieth century, Greater Adjutants were said to have numbered in the millions. They were common sightings in northern India, Bangladesh, Nepal and southern Vietnam. Beginning in the mid-1980s the population began to decline sharply.[14] Out of its total projected global population of 1,000 (record low by 1990s) about 600–700 birds are found in Assam. 'Ignored' by people, the 'gauche' Hargila, living off garbage dumps, with a bald head and an enormous pouch on its neck, has generally been considered a bad omen and a pest. In local parlance, it is also used to refer to lanky people. With such a despondent association with human society, the species had been almost rendered extinct.

Generally, pollution of freshwater systems, felling of large nesting trees, and discarding of human corpses in public waste dumps are all thought to contribute to the swift loss of this species. In Assam, a number of reports of disease in this species also seem to contribute to its rapid vanishing. In a survey carried out in the state, it was revealed that only 30 per cent of the residents had the knowledge that Greater Adjutants are endangered. This tells us about the pathetic condition of eco-awareness among the masses and that larger community awareness may help recovering this unique species.[15]

With the arrival of Purnima Devi Barman, associated with the wildlife NGO Aaranyak, this despondency transformed into a tale of heroic reclamation. As a child, Purnima would watch the birds in the paddy fields of her grandmother's village of Pubmajir in the Kamrup district of Assam.[16] In 2007, she selected *L. dubius* as the subject for her PhD in wildlife biology. As a part of her fieldwork, she worked in the villages of Kamrup—one of the few places on earth where this species still exists. Barman found very few Greater Adjutant nests. What she found in abundance were people who viewed the birds as dirty pests, to the extent that landowners habitually cut down the tall trees that Greater Adjutants nested in. Naturally, with big birds like who produce prodigious amount of fecal matter, who could blame the landowners? The birds gulp down anything, dead or alive, and leave behind a mess. This 'mess', coupled with a lack of awareness of the role the bird plays in the ecosystem—they are important scavengers of unwanted human waste, including unburied corpses as well as other sizable carrion, and contribute to sanitation and disease control–their tendency to nest near human settlements have proved near fatal to them.[17]

Barman understood that unlike other 'glamorous' species of animals like tigers or rhinoceros that have a garnered global popularity, the Greater Adjutant never enjoyed the same kind of appreciation. Thus, she took it upon herself to save the bird and ended up forming a group of village women dedicated to the cause of protecting and conserving the vanishing 'ugly' bird. An affectionate epithet 'Hargila *baideo*' or 'stork sister' has been bestowed on Barman.[18]

She embarked on a what was first a one-woman campaign, and was then eventually joined by several comrade sisters, to infuse a sense of local community pride and 'ownership' in the Greater Adjutant. Her line of argument with villagers was:

> I have two daughters. If they make my house messy, should I dislike them? What do I do? I clean it. Why? Because they are my daughters! I have ownership; I'm their guardian. You are the guardian of this stork, and these birds are making you very proud. Your name is going to be famous throughout the world because if you protect the bird, then all will salute you.[19]

She started off with traditional approaches to mobilize public awareness—praising the bird and its ecosystem with the help of slideshows. However, she soon realized the limited nature of these methods and knew that she needed to do more. Educating children of landowners was one substantial move taken by Barman. On one occasion, to the surprise of many, she also managed to get a popular India movie star on board, with the intention to garner public interest towards the cause; this did have positive results, as she exclaims.[20] Barman also employed the game called 'web-of-life string' that saw even the shyest of female villagers participating. The women were particularly surprised, and proud, to learn that this very rare bird had chosen to breed only in local villages.[21]

Since 2010, positive reports have started to pour in, stating that not a single tree with Greater Adjutant nests has been cut down in the areas of her outreach work. In 2015, Barman counted up to 150 nests against around 30 in 2007. In 2016, she estimates that there around 550 nests in Kamrup district.[22]

Barman's relentless efforts reiterate the point that winning is not just about leading the way; it also humours patience and resources. In the long-term view, recognizing success is not a matter of months, or even years, but might involve decades of untiring efforts.

Such stories succeed because of remarkable people, some just like us. They are considered extraordinary because they imagined ways to reach their goals, worked out how to tread certain paths, established relationships with people, and motivated others to follow their lead.

Not everyone who has the means uses it to plan the cumulative good of all living beings on earth. I consider politicians to be one such category; though they have all the state resources at their disposal and a gamut of planners to make use of all those resources to further sustainable growth, seldom does one comes across politicians who have an intent and grit to work for 'others' or to take care of the biosphere which sustains her/him. Having said that, it is not entirely a disconsolate picture; we have some political leaders, even though the numbers are very shy of producing a laudable portent, who have tried to make a difference. I could perhaps talk of Theodore Roosevelt here. Often considered a 'conservation president', Roosevelt is one classic example of hunter-turned-conservationist.

As a child with an intense fascination for 'teddy' bears, I had heard about a person who participated in the 'loathsome' activity of hunting a bear and ended up not killing it. The ending made me immensely happy. As I grew up, I read about Roosevelt's life. He is honoured with a national park that bears his name and carries his memory. His writings, which are an illustration of copious hunting stories, are also laced with requiems for the loss of habitat and species. It is noteworthy that gradually conservation became one of Roosevelt's foremost concerns. The extermination of elk, the obliteration of bison, deer, bighorn sheep, and other game species was a loss which Roosevelt felt was suggestive of society's insight about our natural resources. He understood the effects of over-grazing and that natural resources are exhaustible. He wrote:

> We have become great because of the lavish use of our resources. But the time has come to inquire seriously what will happen when our forests are gone, when the coal, the iron, the oil, and the gas are exhausted, when the soils have still further impoverished and washed into the streams, polluting the rivers, denuding the fields and obstructing navigation.[23]

Roosevelt dedicated his later life to reduce humankind's 'wanton vandalism' of nature. Sadly, the effects of this destruction lingers, even as good Samaritans do not lose hope and continue to save Mother Nature.

The story of the California Condor (*Gymnogyps californianus*) is one of optimism, even though it is still listed as Critically Endangered[24] by the IUCN. In the twentieth century, numbers of this North American land bird dropped so precipitously that only twenty-seven individuals were left by 1987, at which point all were taken into captivity to try to save the species. Stuart Butchart states: 'They are basically a vulture.

They feed on carcasses and ingest fragments of lead shot, and because they live for decades, that can accumulate over time. It's incredibly poisonous.'[25] They were also plagued by other problems like ingestion of rubbish that included hazardous glass, crashing against electrical pylons, and the consumption of the banned (since 1972) insecticide DDT that resulted in thinning of eggs.[26]

Nevertheless, a relentless captive breeding programme resulted in success and in 1991 the condors were released into the wild. Now there are ninety-three mature individuals among a population of 300 birds in the wild.[27] Perhaps, the status of this bird will soon improve by 2024.

The story of tigers is one of 'decline of one of Earth's largest predators', said Stuart Chapman at the WWF-UK conservation body meeting.[28] In 1900, the population of tigers in the wild was about 100,000 individuals, with a majority of them found in India.[29] By the early 1970s, rampant poaching and habitat loss let the numbers to drop below 500 individuals worldwide.[30] Much to the relief of the species, Project Tiger was launched in India in 1973, resulting in the creation of fifty reserves spread across 40,000 sq. km., and an increase in population from 268 individuals in 1973 2,967 tigers now.[31] There was a monumental 6 per cent per annum rise in population from 2006 to 2018 in India. Even though a number of champions led us to this success, Shri Kailash Sankhala, the first director of Project Tiger, needs special mention, as he turned the tide for wildlife conservation in the country, particularly of tigers. He was the pioneering Indian field officer who undertook tangible field research in understanding tiger behaviour, population dynamics, and daily activities.[32]

It is always a visionary and passionate individual who foresees a different world and pleas for its conservation. Independent, resilient thinkers see their way around apparently obdurate problems. Felix Finkbeiner,[33] a young German environmentalist and founder of the international organization 'Plant-for-the-Planet' has been relentlessly trying to make the earth a greener place since the age of 9, after being inspired by the lecture of Nobel laureate Wangari Maathai. In 2011, at the United Nations General Assembly, the 9-year-old Finkbeiner spoke about the inaction on the part of adults who were supposed to preserve the planet for future generations. His primary thought is the stark reality of our times: 'For most adults, it's an academic question. For many of us children, it's a question of survival', he said. 'Twenty-one hundred is still in our lifetime.'[34]

Grasping that he and his generation 'cannot trust that adults'[35] and would have to pay for their passivity, Finkbeiner established the environmental group Plant-for-the-Planet, which eventually partnered with the UN's Billion Tree campaign. Together, they have set a goal of planting one trillion trees—approximately 150 trees for each person on Earth.[36] Finkbeiner has planted more than 14 billion trees so far.[37]

In another part of the world, someone else was also 'planting' a hopeful future for younger generations. A 65-year-old Indian, Jadav Molai Payeng,[38] a 'simple villager' from Majuli, north-east India, has been single-handedly planting trees since 1979 and has now turned 550 hectares (over 1300 acres) of barren land into a biodiversity hotspot, a lush green forest[39] dubbed the 'Molai forest'. This hotstpot now attracts 80 per cent of world's migratory birds, claim wildlife experts.[40]

Majuli, the world's largest river island, has been threatened with erosion by the river Brahmaputra, leading it to almost losing its title of being the 'largest'. The island houses close to 150,000 people who have lost more than half of their home's (Majuli) landmass since 1917.[41] Payeng has been planting tree for thirty years now, and is firmly resolute about continuing doing so.

People often anthromorphize animals. While this receives its share of criticism, people who have a mystical connect with the non-human world claim that these are not emotional fanaticisms but have scientific basis[42] as animals have sentience— they feel, think, express. Biology professor Nathan Lents, in his book *Not So Different: Finding Human Nature in Animals* explains that animals regularly fall in love, delineate rules for fair play, hold funerals, communicate with an extensive vocabulary, etc. Elephants grieve, monkeys fight against inequality, wolves miss their partners, and prairie dogs end up naming the humans that they come across.

In this context, Mishak Nzimbi,[43] a night watchman turned elephant keeper from Nairobi Elephant Nursery stands out. He is known for his heart and is recognized by every elephant that he has ever met. Mishak was chosen by more than 130 orphaned elephants as their family. He workd miracles with heartbroken baby elephants who do not want to live. He radiates a unique 'confidence, humility and a depth of understanding of elephants'[44] that has turned things around overnight for these animals on several occasions. Mishak is a true elephant whisperer and must have a pure heart 'because all elephants can read one's heart and over 130 elephants simply cannot be wrong!'[45]

Loving the non-human world is uplifting and nourishing for the soul. I fail to fathom the hubris of humans that continues to wage war on wildlife even at the cost of their own extinction. Conservation champions have time and again shown us that conservation is not about resolutions on papers but a resolute mindset with longer and the wider view—horizontally, that benefits humans and non-humans alike, and vertically, that covers all future generations. Sadly, most (if not all) governments function keeping their transient political reigns in mind, and environmental issues are reflected in their policies as long as they provide revenue in material terms. However, if nature matters to us, however limited in number we might be, we should let our leaders know. We have a role, and an important one at that, to play. Speak out for the silent members of our planet.

People can help in various ways—volunteer for hands-on activities like restoring habitats, surveying local species, fundraising for NGOs, donating money, etc. Go out of your way to educate people in your vicinity, and market your ideas and views through social media platforms. If you see a problem around you and no one else is tackling it, start your own campaign, be consistent and rigorous; people will join in.

Most success stories across the world have an ambitious and an enterprising mind behind them. However, with ambition comes the possibility of failure. Having said that, I believe it's worth fighting for and failing (if at all), than doing nothing and letting this planet be destroyed. The larger the scale of threats to nature, the bigger our ambitions must be to save it. One small step can lead to a big change in the world; we only have to take that one small stride. With trees and animals gone, humans will be gone soon too. No progress is worth this.

NOTES

1. Andrew Balmford, *Wild Hope: On the Frontlines of Conservation Success*, Chicago: U of Chicago Press, 2012.
2. Marlee Kokotovic, 'The Bee is Oficially on the Verge of Extinction', 2019, see https://www.nationofchange.org/2019/12/05/the-bee-is-officially-on-the-verge-of-extinction/, accessed 4 April 2021.
3. 'Frightening Number of Plant Extinctions Found in Global Survey',*The Guardian*, 10 June 2019.
4. Ibid.
5. E.S. Brondizio et al., *Global Assessment Report on Biodiversity and Ecosystem Services of the Intergovernmental Science-Policy Platform on Biodiversity and Ecosystem Services*, Bonn, Germany: IPBES Secretariat, 2019.
6. Andrew Balmford, *Wild Hope: On the Frontlines of Conservation Success*, Chicago: University of Chicago Press, 2012, p. 5.
7. 'Maguri Beel', *eBird* (website), see https://ebird.org/hotspot/L2525751, accessed 4 January 2022.
8. 'Maguri Beel', *Tourginie* (website), see https://www.tourgenie.com/

discover/discover-assam/maguri-beel-sightseeing, accessed 5 November 2021.

9. Anupam Chakravarty, 'NGT Committee Renders Entire Baghjan Oil Field Illegal'. *Down to Earth*, 3 November 2020, see https://www.downtoearth.org.in/news/environment/ngt-committee-renders-entire-baghjan-oil-field-illegal-74077#:~:text=Published%3A%20Tuesday%2003%20November%202020&text=The%20Baghjan%20oilfield%2C%20along%20with,NGT)%20committee%20said%20November%203, accessed 6 November 2020.
10. 'Baghjan Blowout Well Sealed Fully', *Hindu*, 16 November 2020, see https://www.thehindu.com/news/national/other-states/baghjan-blowout-well-sealed-fully/article33103989.ece, accessed 4 July 2021.
11. Tora Agarwala, 'Assam Gas Leak: Why it's Tough to Plug, and What Threat It Poses to the Area', *Indian Express*, 8 June 2020. Retrieved 8 June 2020.
12. Suzana Camargo, 'Fires in the Amazon Have Already Impacted 90% of Plant and Animal Species', *Mongabay* (website), 2021 see https://news.mongabay.com/2021/09/fires-in-the-amazon-have-already-impacted-more-than-90-of-animal-and-plant-species-in-the-biome/, accessed 7 November 2021.
13. X. Feng et al., 'How Deregulation, Drought and Increasing Fire Impact Amazonian Biodiversity', *Nature*, vol. 597, 2021, pp. 516–521.
14. Christina Kwapich, 'Leptoptilos dubiusgreater Adjutant', Animal Diversity Web (website), 2011, see https://animaldiversity.org/accounts/Leptoptilos_dubius/#ecosystem_roles, accessed 20 December 2022.
15. A. Elliot, J. Del-Hoyo and D. Christie, *Handbook of the Birds of theWorld*, vol 1: Ostrich to Ducks, Barcelona, Spain: Lynx Edicions, 1992.
16. Cat Kutz, 'Building An All-Women Army for Conservation In India: Earth Optimism Speaker Interview with Purnima Devi Barman', *Smithsonian Magazine*, 2020, see https://www.smithsonianmag.com/blogs/earth-optimism/2020/03/06/building-all-woman-army-conservation-india/, accessed 7 April 2020.
17. Ibid.
18. Ibid.
19. Diane Toomey, 'From Loathed to Loved: Villagers Rally to Save Greater Adjutant Stork', *Mongabay* (website), 2016, see https://news.mongabay.com/2016/12/from-loathed-to-loved-villagers-rally-to-save-greater-adjutant-stork/, accessed 2 January 2021.
20. Ibid.

21. Ibid.
22. Ibid.
23. 'Theodore Roosevelt and Conservation', *National Park Service* (website), 2017, see https://www.nps.gov/thro/learn/historyculture/theodore-roosevelt-and-conservation.htm, accessed 7 April 2021.
24. BirdLife International, 'Gymnogyps californianus', *IUCN Red List of Threatened Species* (website), 2020 see https://dx.doi.org/10.2305/IUCN.UK.2020-3.RLTS.T22697636A181151405.en, accessed 7 February 2022.
25. Shelley Singh, 'Some Successful Animal Conservation Projects around the World', *Economic Times*, 2020, see https://economictimes.indiatimes.com/industry/miscellaneous/some-successful-animal-conservation-projects-around-the-world/tiger/slideshow/77312047.cms, accessed 14 January 2021.
26. Adam Vaughan, 'Ten Conservation Success Stories when Species Came Back from the Brink', *News Scientist*, 2021, see https://www.newscientist.com/article/mg24933223-400-ten-conservation-success-stories-when-species-came-back-from-the-brink/#ixzz7KBttwW6F, accessed 7 February 2022.
27. Singh, 'Some Successful Animal Conservation Projects around the World'.
28. Ibid.
29. Rinchen Norbu Wangchuk, 'Guilt-Stricken on Killing a Tiger, Padma Shri Officer Went On To Change Their Destiny in India', *The Better India*, 2020, see https://www.thebetterindia.com/223419/padma-shri-kailash-sankhala-project-tiger-kanha-sariska-india-unsung-hero-nor41/, accessed 2 February 2022.
30. Ibid.
31. Ibid.
32. Heather Brady, '5 Times People Used Trees to Change the World', *National Geographic*, 2018, see https://www.nationalgeographic.com/culture/article/trees-impact-india-london-georgia-erosion-forest-culture, accessed 21 May 2020.
33. M. Parabalika Borah, 'The Man who Built a Forest: Jadav Payeng', *The Hindu*, 3 December 2020, see https://www.thehindu.com/society/history-and-culture/jadav-payeng-tells-how-he-became-the-forest-man-of-india/article33230503.ece, accessed 2 June 2021.
34. Ibid.
35. Ibid.
36. Ibid.
37. Ibid.

38. Ibid.
39. 'Our Elephant Cherub and the Elephant Whisperer', *Sheldrick Wildlife Trust* (website), 2021, see https://www.sheldrickwildlifetrust.org/news/updates/our-elephant-cherub-and-the-elephant-whisperer, accessed 20 December 2021.
40. Ibid.
41. Borah, 'The Man who Built a Forest'.
42. Nathan H. Lents, *Not So Different: Finding Human Nature in Animals*, New York: Columbia University Press, 2016.
43. 'Our Elephant Cherub and the Elephant Whisperer'.
44. Ibid.
45. Ibid.

Bibliography

'Amur Leopard the Beautiful', *Luc Vertebrates* (blog), 2014, see https://lucvertebrates.wordpress.com/2016/12/09/amur-leopard-the-beautiful/, accessed 14 December 2014.

'Amur Leopard: Probably the World's Rarest Cat?', *WWF* (website), n.d.

'Amur Leopard', *North-East Asian Subregional Programme for Environmental Cooperation* (website), n.d., see http://neaspec.org/sites/default/files/Fact%20Sheet%20of%20Amur%20Leopard.pdf, accessed 14 December 2020.

'Amur Leopard', *WWF* (website), n.d., see https://wwf.panda.org/knowledge_hub/endangered_species/amur_leopard2/, accessed 14 December 2020.

'Appendices', *Convention on International Trade in Endangered Species of Wild Flora and Fauna* (website), 2017, see https://cites.org/eng/app/appendices.php, accessed 23 April 2019.

'*Arenga obtusifolia* Palm', Global Invasive Species Database (website), see http://www.issg.org/database/species/ecology.asp?si=1834&fr=1&sts&sts=&lang=EN, accessed 17 February 2017.

'Baghjan Blowout Well Sealed Fully', *The Hindu*, 16 November 2020, see https://www.thehindu.com/news/national/

other-states/baghjan-blowout-well-sealed-fully/article33103989.ece, accessed 4 July 2021.

'Bamboo Flowering', *Bamboo Plants and Anatomy* (website), 2019 see http://www.bamboobotanicals.ca/html/about-bamboo/bamboo-flowering.html, accessed 13 June 2020.

'Biodiversity', *National Geographic*, n.d., see https://education.nationalgeographic.org/resource/biodiversity, accessed 1 January 2022.

'Birds of the Bible: Solomon's Wisdom about Birds', *Lee's Birdwatching Adventures* (blog), 2013, see https://leesbird.com/2013/11/12/birds-of-the-bible-solomons-wisdom-about-birds-i/, accessed 20 December 2022.

'CITES Amendments', *CITES* (website), n.d., see https://cites.org/sites/default/files/eng/cop/08/prop/E08-Prop-EQ1_to_EQ5_Panthera.PDF, accessed 14 December 2020.

'Endangered Species', *WWF* (website), n.d., see https://wwf.panda.org/knowledge_hub/endangered_species/amur_leopard2/, accessed 14 December 2020.

'Frightening Number of Plant Extinctions Found in Global Survey', *The Guardian*, 10 June 2019.

'Houston Toad (Anaxyrus houstonensis)', *Texas Parks and Wildlife* (website), 4 April 2021, see https://tpwd.texas.gov/huntwild/wild/species/htoad/, accessed 20 December 2022.

'Invasive Species', *National Wildlife Federation* (website), n.d., see https://www.nwf.org/Educational-Resources/Wildlife-Guide/Threats-to-Wildlife/Invasive-Species, accessed 5 March 2020.

'Maguri Beel', *eBird* (website), see https://ebird.org/hotspot/L2525751, accessed 4 January 2022.

'Maguri Beel', *Tourginie* (website), see https://www.tourgenie.com/discover/discover-assam/maguri-beel-sightseeing, accessed 5 November 2021.

'Multi-species Action Plan to Conserve African-Eurasian

Vultures', *IUCN* (website), see https://portals.iucn.org/library/sites/library/files/documents/2017-064.pdf, accessed 20 December 2022.

'Our Elephant Cherub and the Elephant Whisperer', *Sheldrick Wildlife Trust* (website), 2021, see https://www.sheldrickwildlifetrust.org/news/updates/our-elephant-cherub-and-the-elephant-whisperer, accessed 20 December 2021.

'Peregrine Falcon (Falco peregrinus)', *U.S. Fish and Wildlife Service* (website), 2006, see https://ecos.fws.gov/ecp/species/8483, accessed 20 December 2022.

'Priority Species', *WWF* (website), n.d., see https://wwf.panda.org/knowledge_hub/endangered_species/ , accessed 14 December 2020.

'Probably the World's Rarest Cat?', *WWF* (website), n.d., see https://www.wwf.org.uk/learn/wildlife/amur-leopards, accessed 14 December 2020.

'Radio–Telemetry Study of Elephants in Buxa Tiger Reserve and Adjoining Areas in North West Bengal', Asian Nature Conservation Foundation (website), 2013, see http://www.asiannature.org/our-programmes/asian-elephant-research-and-conservation-aerc/human-elephant-conflict-mitigation-pr-2, accessed 15 December 2022; S. Chakraborty, 2015. Human-animal conflicts in northern West Bengal: losses on both sides', *Int J Pure App Biosci*, vol. 3, 2015, pp. 35–44.

'Species Factsheet: *Rhinoptilus bitorquatus*', *BirdLife International* (website), 2020, see http://www.birdlife.org, accessed 12 May 2020.

'Species Factsheet: *Ardea insignis*', *BirdLife International* (website), 2020, see http://www.birdlife.org, accessed 2 June 2020.

'State of India's Birds factsheet: Yellow-throated Bulbul, Pycnonotus xantholaemus', *State of India's Birds* (website), 2020, see https://www.stateofindiasbirds.in/species/yetbul1/, accessed 6 July 2020.

'Strategy and Action Plan for the Conservation of Rhinos in Indonesia', Directorate General of Forest Protection and Nature Conservation; Ministry of Forestry of the Republic of Indonesia, 2007.

'Theodore Roosevelt and Conservation', *National Park Service* (website), 2017, see https://www.nps.gov/thro/learn/historyculture/theodore-roosevelt-and-conservation.htm, accessed 7 April 2021.

'Why Save Endangered Species?', *U.S. Fish and Wildlife Service* (website), 4 April 2021, see https://www.fws.gov/sites/default/files/documents/Why-Save-Endangered-Species-Brochure_0.pdf, accessed 20 December 2022.

'Wildlife Conservation Threats', *WCS Russia* (website), n.d., see https://russia.wcs.org/en-us/Wildlife/Amur-Leopards/Conservation-Threats.aspx, accessed 14 December 2020.

'Wildlife Conservation Threats', *WWF* (website), n.d., see https://russia.wcs.org/en-us/Wildlife/Amur-Leopards/Conservation-Threats.aspx, accessed 14 December 2020.

'Wildlife Ecology', *WCS Russia* (website), n.d., see https://russia.wcs.org/en-us/Wildlife/Amur-Leopards/Ecology.aspx., accessed 14 December 2020.

'Wolves in Yellowstone: A Short History', *Yellowstone Insider*, see https://yellowstoneinsider.com/2009/05/03/wolves-in-yellowstone-a-short-history/, accessed 8 December 2022.

Acharja, I., 'Evaluation of Nest Habitat, Site Preferences and Architecture of the Critically Endangered White-bellied Heron *Ardea insignis* in Bhutan', *Bird Conservation International*, 2019, pp. 1–19.

Agarwala, Tora, 'Assam Gas Leak: Why it's Tough to Plug, and What Threat It Poses to the Area', *Indian Express*, 8 June 2020, Retrieved 8 June 2020.

Aggarwal, Mayank, 'India's Elephant Population Decreases by 10% to 27,312', *LiveMint*, 12 August 2017, see https://www.livemint.com/Politics/OhPuqM871gkivaj5zetCZJ/Indias-elephant-population-decreases-by-10-to-27312.html, accessed 15 December 2022.

Ahmed, M.F. and A. Das, 'Tortoises and Freshwater Turtles of Kaziranga National Park, Assam: Diversity, Distribution, Conservation Status', in *Freshwater Turtles and Tortoises of India*, ENVIS Bulletin, vol. 12, no. 1, 2009, pp. 57–70.

Ahsan, M.F. and M.A. Saeed, 'The Bostami Turtle, Trionyx nigricans Anderson: Population Status, Distribution, Historical Background and Length-weight Relationship', *Bombay Natural History Society*, vol. 86, 1989, pp. 2–5.

Ahsan, M.F., M.N. Haque and C.M. Fugler, 'Observations on Aspideretes nigricans, a Semi-domesticated Endemic Turtle from Eastern Bangladesh', *Amphibia-Reptilia*, vol. 12, 1991, pp. 131–6.

Ali, S. and A.R. Rahmani, 'Study of Ecology of Certain Endangered Species of Wildlife and their Habitats: The Great Indian Bustard', 2nd Annual Report, Bombay: Bombay Natural History Society, 1982.

Ali, S. and S. Dillon Ripley, *Compact Handbook of the Birds of India and Pakistan*, 2nd edn., Oxford: Oxford University Press, 1989.

Ali, S. and S.D. Ripley, *Handbook of the Birds of India and Pakistan Together with Those of Nepal, Sikkim, Bhutan and Ceylon*, New Delhi: Oxford University Press, 1983.

Ali, S. and S.D. Ripley, 'Handbook of the Birds of India and Pakistan', vol. II, Delhi: Oxford University Press, 1969.

Ali, S., 'The Moghul Emperors of India as Naturalists and Sporstmen. Part II', *Journal of the Bombay Natural History Society*, vol. 32, no. 1, 1927, pp. 34–63.

Ali, S., 'The Moghul Emperors of India as Naturalists and Sportsmen. Part II', *Journal of the Bombay Natural History Society*, vol. 32, no. 1, 1927, pp. 34–63.

Ali, S., 'The Moghul Emperors of India as Naturalists and Sportsmen', *Journal of the Bombay Natural History Society*, vol. 31, 1927, pp. 833–61.

Anderson, D.M., M.L. Richlen and K.A. Lefebvre, 'Harmful Algal Blooms in the Arctic', *Arctic Report Card*, 2018, see https://www.arctic.noaa.gov/Report-Card, accessed 19 December 2022.

Annandale, N. and M. Shastri, 'Relics of the Worship of Mud Turtles (Trionychidae) in India and Burma', *J. Proc. Asiatic Soc. Bengal* (n.s.), vol. 10, 1914, pp. 131–8.

Antram, C.B., *Butterflies of India*, Calcutta and Simla: Thacker, Spink & Co., 1924.

Appelman, F.J., 'Rhinoceros sondaicus in West-Java', *Tropische Natuur*, vol. 23, 1934, pp. 73–6.

'Ardeotis nigriceps', IUCN Red List of Threatened Species (website), 2018, see http://dx.doi.org/10.2305/IUCN.UK.2018-2.RLTS.T22691932A134188105.en, accessed 20 December 2022.

Ashby, K.R. and C. Santiapillai, 'The Ecology of Free Living Water Buffalo (*Bubalus bubalis*) in Sri Lanka with Particular Reference to Ruhuna National Park', *Tigerpaper*, vol. 14, no. 3, 1987, pp. 20–6.

Asmat, G.S.M, 'Nilssonia nigricans', *IUCN Red List of Threatened Species* (website), 2002, see https://www.iucnredlist.org/fr/species/2173/2778172, accessed 20 December 2022.

Athreya, V. et al., 'Big Cats in our Backyards: Persistence of Large Carnivores in a Human Dominated Landscape in India', *PLoS One*, vol. 8, no. 3, 2013; P.S. Bindra, *The Vanishing: India's Wildlife Crisis*, Delhi: Penguin, 2017.

Atwood, T.C. et al., 'Forecasting the Relative Influence of Environmental and Anthropogenic Stressors on Polar Bears', *Ecosphere*, vol. 7, 2016.

Babur, Zahiru'd-din Muhammad, *Babur-nama* (Memoirs of Babur), tr. Annette Susannah Beveridge, vol. II, London: Luzac & Co, 1922, p. 46.

Bagley, M., 'Triassic Period Facts: Climate, Animals and Plants', *Live Science* (website), 2014, see https://www.livescience.com/43295-triassic-period.html#:~:text=The%20Triassic%20Period%20was%20the,the%20oceans%20began%20to%20diversify, accessed 20 December 2022.

Baillie, J.E.M. et al., 'Manis crassicaudata', *The IUCN Red List of Threatened Species 2014* (website), 2016, see https://www.

iucnredlist.org/species/12761/123583998, accessed 28 October 2016.

Baker, E.C.S, *The Nidification of birds of the Indian Empire*, London: Taylor and Francis, 1929.

Baker, Harry, 'A Secret Polar Bear Population Has Been Found in an "Impossible" Location', *Science Alert* (website), 2022, see https://www.sciencealert.com/polar-bears-discovered-secretly-living-in-seemingly-inhospitable-habitat, accessed 19 December 2022.

Balmford, Andrew, *Wild Hope: On the Frontlines of Conservation Success*, Chicago: University of Chicago Press, 2012.

Bellard, Céline et al., 'Impacts of Climate Change on the Future of Biodiversity', *Ecology Letters*, vol. 15, no. 4, 2012.

Bemment, N., 'Orang utan EEP Best Practice Guidelines', vol. 1, 1st edn., ed. Neil Bemment, Barcelona: Great Ape Taxon Advisory Group, 2019, https://www.eaza.net/assets/Uploads/CCC/OU-EEP-Best-Practice-Guidelines-final.pdf, accessed 20 December 2022.

Berardelli, Jeff, 'With More Species at Risk of Extinction, Study Warns of "Biological Annihilation"', *CBS News*, 2020, see https://www.cbsnews.com/news/species-extinction-risk-biological-annihilation-study/, accessed 20 December 2022.

Bhardwaj, G.S. et al., 'Role of Herbaceous Vegetation in Habitat Utilization by Critically Endangered Great Indian Bustard Ardeotis nigriceps (Vigors) in the Indian Thar Desert', *Indian Forester*, vol. 143, 2017, pp. 975–84.

Bhushan, B. and A.R. Rahmani, 'Food and Feeding Behaviour of the Great Britain Bustard Ardeotis nigriceps (Vigors)', *Journal of the Bombay Natural History Society*, vol. 89, 1992, pp. 27–40.

Bhushan, B., 'Rediscovery of the Jerdon's or Double-banded Courser *Cursorius bitorquatus*', *Journal of the Bombay Natural History Society*, vol. 83, 1986, pp. 1–14.

BirdLife International, 'Asarcornis scutulata', *The IUCN Red List of Threatened Species* (website), 2017, see http://dx.doi.

org/10.2305/IUCN.UK.2017-1.RLTS.T22680064A110103586.en, accessed 17 December 2022.

BirdLife International, 'Species Factsheet: *Pycnonotus xantholaemus*', *Data Zone* (website), 2020, see http://datazone.birdlife.org/species factsheet/22712719 on 02/09/2020, accessed 17 December 2022.

———, *Threatened Birds of Asia: The BirdLife International Red Data*, Cambridge: BirdLife International, 2001.

———, *Threatened Birds of Asia: The BirdLife International Red Data Book*, Cambridge: Birdlife International, 2001.

———, 'Gymnogyps californianus', *IUCN Red List of Threatened Species* (website), 2020 see https://dx.doi.org/10.2305/IUCN.UK.2020-3.RLTS.T22697636A181151405.en, accessed 7 February 2022.

Bist, S.S., 'An Overview of Elephant Conservation in India', *Indian Forester*, vol. 128, no. 2, 2002.

Blake, S. et al., 'Forest Elephants: Tree Planters of the Congo', *Biotropica*, vol. 41, no. 4, 2009, pp. 459–68; Jeremy Hance, 'Elephants: The Gardeners of Asia's and Africa's forests', *Mongabay* (website), 25 April 2011, see https://news.mongabay.com/2011/04/elephants-the-gardeners-of-asias-and-africas-forests/, accessed 15 December 2022.

Blanford, *The Fauna of British India*, vol. IV, London: Taylor and Francis, 1868.

Blanford, W.T., *The Fauna of British India, including Ceylon and Burma*, vol. III, London: Taylor and Francis, 1895.

Blyth, E. 'Report of Curator, Zoological Department', *J. Asiatic Society of Bengal*, vol. XVII, 1848, p. 254.

Book of Genesis, see http://www.vatican.va/archive/bible/genesis/documents/bible_genesis_en.html, accessed 20 December 2022.

Borah, M. Parabalika, 'The Man who Built a Forest: Jadav Payeng', *The Hindu*, 3 December 2020, see https://www.thehindu.com/society/history-and-culture/jadav-payeng-tells-how-he-became-the-forest-man-of-india/article33230503.ece, accessed 2 June 2021.

Bourns, Timothy, 'The Language of Birds in Old Norse Tradition', *Journal of English and German Philology*, vol. 12, no. 2, pp. 209–28.

Bradshaw, G.A. et al., 'Elephant Breakdown', *Nature*, vol. 433, no. 7028, 2005, p. 807.

Brady, Heather, '5 Times People Used Trees to Change the World', *National Geographic*, 2018, see https://www.nationalgeographic.com/culture/article/trees-impact-india-london-georgia-erosion-forest-culture, accessed 21 May 2020.

Broders, O., T. Osborne and M.A. Wink, 'Mtdna Phylogeny of Bustards (Family Otididae) Based on Nucleotide Sequences of the Cytochromeb-gene', *Journal für Ornithologie*, vol. 144, 2003, pp. 176–85.

Brondizio, E.S. et al., *Global Assessment Report on Biodiversity and Ecosystem Services of the Intergovernmental Science-Policy Platform on Biodiversity and Ecosystem Services*, Bonn, Germany: IPBES Secretariat, 2019.

Bucher E.H., 'The Causes of Extinction of the Passenger Pigeon', in *Current Ornithology*, vol. 9, ed. D.M. Power, Boston: Springer, 1992.

Camargo, Suzana, 'Fires in the Amazon Have Already Impacted 90% of Plant and Animal Species', *Mongabay* (website), 2021 see https://news.mongabay.com/2021/09/fires-in-the-amazon-have-already-impacted-more-than-90-of-animal-and-plant-species-in-the-biome/, accessed 7 November 2021.

Cannon, J., 'Orangutan Culture in Focus in "Person of the Forest": Q&A with Researchers Cheryl Knott and Robert Rodriguez Suro', Mongabay (website), 23 February 2018, see https://news.mongabay.com/2018/02/orangutan-culture-in-focus-in-person-of-the-forest-qa-with-researchers-cheryl-knott-and-robert-rodriguez-suro/, accessed 20 December 2022.

Carrington, D., 'Earth's Sixth Mass Extinction Event under Way, Scientists Warn', *The Guardian*, 2017, see https://

www.theguardian.com/environment/2017/jul/10/earths-sixth-mass-extinction-event-already-underway-scientists-warn, accessed 20 December 2022.

Carson, M. and G. Peterson, eds., *Arctic Resilience Report*, Stockholm: Stockholm Environment Institute and Stockholm Resilience Centre, 2016.

Chakravarty, Anupam, 'NGT Committee Renders Entire Baghjan Oil Field Illegal', *Down to Earth*, 3 November 2020, see https://www.downtoearth.org.in/news/environment/ngt-committee-renders-entire-baghjan-oil-field-illegal-74077#:~:text=Published%3A%20Tuesday%2003%20November%202020&text=The%20Baghjan%20oilfield%2C%20along%20with,NGT)%20committee%20said%20November%203, accessed 6 November 2020.

Challender, D.W., C. Waterman and J.E. Baillie, *Scaling up Pangolin Conservation. IUCN SSC, Pangolin Specialist Group Conservation Action Plan*, London: Zoological Society of London, 2014.

Challender, D.W.S., 'Asian Pangolins: Increasing Affluence Driving Hunting Pressure', https://www.wwfindia.org/about_wwf/enablers/traffic/media/?uNewsID=17061, accessed 7 November 2019.

Chatterjee, B., 'India has 10% of World's 50 Most Threatened Freshwater Turtle Species: Report', *Hindustan Times*, 2018, see https://www.hindustantimes.com/mumbai-news/india-has-10-of-world-s-50-most-threatened-freshwater-Turtle-species-report/story-srgbMsjH2lp9jxjTTXK1XL.html, accessed 20 December 2022.

Cheke, A. et al., 'Did the Dodo Do It?', *Animal Kingdom*, vol. 87, 1984, pp. 4–6.

Cheke, A., 'The Legacy of the Dodo: Conservation in Mauritius', *Oryx*, vol. 21, no. 1, 1987, pp. 29–36.

Cheke, Anthony, 'Extinct Birds of the Mascarenes and Seychelles: A Review of the Causes of Extinction in the Light of an Important New Publication on Extinct Birds', *Phelsuma*, vol. 21, 2013, pp. 4–19.

Choo, S.W. et al., 'Pangolin Genomes and the Evolution of Mammalian Scales and Immunity', *Genome Research*, vol. 26, no. 10, 2016, pp. 1312–22.

Choudhury, A.U., 'Occurrence of the Wild Water Buffalo *Bubalus arnee* in Mizoram and Manipur', *Journal of the Bombay Natural History Society*, vol. 99, no. 3, 2002, pp. 512–16.

———, 'On the Occurrence of the Wild Water Buffalo *Bubalus arnee* in the Barak Valley Districts of Assam', *Journal of the Bombay Natural History Society*, vol. 98, no. 2, 2001, pp. 270–1.

———, 'Saving the Wild Buffalo', *The India Magazine*, September 1993, p. 75.

———, 'Status of the Wild Water Buffalo *Bubalus arnee* in Lohit District, Arunachal Pradesh', *Journal of the Bombay Natural History Society*, vol. 96, no. 1, 1999, pp. 130–2.

———, 'The Decline of Wild Water Buffalo in Northeastern India', *Oryx*, vol. 28, no. 1, 1994a, pp. 70–3.

———, 'Wild buffalo *Bubalus bubalis* in Dhakuakhana, Lakhimpur District of Assam', *Journal of the Bombay Natural History Society*, vol. 91, no. 1, 1994, pp. 137–8.

———, 'Wild Water Buffalo *Bubalus arnee*', in *Ecology, Evolution and Behaviour of Wild Cattle: Implications for Conservation*, ed. M. Melletti and J. Burton, Cambridge: Cambridge University Press, 2014, pp. 255–301.

———, 'Wild Water Buffalo *Bubalus bubalis / arnee* in Dibang Valley District of Arunachal Pradesh', *Journal of the Bombay Natural History Society*, vol. 95, no. 1, 1998, pp. 110–12.

CITES/GRASP Orang-utan Technical Mission Indonesia, Geneva: CITES, 2006, see https://cites.org/sites/default/files/common/prog/ape/ID_mission06.pdf, accessed 20 December 2022.

Cockrill, W.R., 'Water buffalo', in *Evolution of Domesticated Animals*, ed. I.L. Mason, London: Longman, 1984, pp. 52–63.

Collins, N.M. and M.G. Morris, *Threatened Swallowtail Butterflies of the World*, The IUCN Red Data Book, Gland and Cambridge: IUCN, 1985.

Conti, M.E., 'Lichens as Bioindicators of Air Pollution', *WIT Transactions on State of the Art in Science and Engineering*, vol. 30, 2008.

Corbet, G.B. and J.E. Hill, *The Mammals of the Indomalayan Region: A Systematic Review*, Oxford: Oxford University Press, 1992.

Cordova, Melanie Greaver, 'Critically Endangered Amur Leopard Faces New Threat', *Cornell Chronicle*, 2018, see https://news.cornell.edu/stories/2018/01/critically-endangered-amur-leopard-faces-new-threat, accessed 14 December 2020.

Crowther, T. et al., 'Mapping Tree Density at a Global Scale', *Nature*, vol. 525, 2015, pp. 201–5.

Cuthbert, Richard et al., 'Continuing mortality of vultures in India associated with illegal veterinary use of diclofenac and a potential threat from nimesulide', *Oryx*, vol. 50, no. 1, 2016.

Cuvier, George, *Essay on the Theory of the Earth*, 1815; repr., Oxford: Oxford University Press, 2019.

Daniel, J.C. and B.R. Grubh, 'The Indian Wild Buffalo, *Bubalus bubalis* (L.) in Peninsular India: A Preliminary Survey', *Journal of the Bombay Natural History*

Das, Niranjan and Sujata Deori, 'Occurrence of White-Winged Wood Duck (*Cairina Scutulata*) in Nameri National Park, Assam, India', *Bird Populations*, vol. 11, 2012, pp. 7–13.

Das, Subhasish et al., 'Conservation Status of Black Soft Shell Turtle (*Nilssonia nigricans*) in Bangladesh', *Journal of Wildlife Research*, vol. 6, no. 2, 2018, pp. 12–16.

Devin, Simon and Jean-Nicolas Beisel, 'Biological and Ecological Characteristics of Invasive Species: A Gammarid Study', *Biological Invasions*, vol. 9, 2007, pp. 13–24.

Dewani, U., 'Forgotten Waterbearers of Guwahati', *India Water Portal* (website), 2016, see indiawaterportal.org/articles/forgotten-water-bearers-guwahati, accessed 20 December 2022.

Dharmakumarsinhji, R., *Study of the Great Indian Bustard, Final Report*, Morges: WWF, 1971.

Díaz, Sandra, Josef Settele and Eduardo Brondízio, 'Summary for Policymakers of the Global Assessment Report on Biodiversity and Ecosystem Services of the Intergovernmental Science-Policy Platform on Biodiversity and Ecosystem Services', 2019, see https://ipbes.net/sites/default/files/downloads/spm_unedited_advance_for_posting_htn.pdf, accessed 20 December 2022.

Dorcas, M.E. et al., 'Severe Mammal Declines Coincide with Proliferation of Invasive Burmese Pythons in Everglades National Park', *Proceedings of the National Academy of Sciences*, vol. 109, no. 7, 2012, pp. 2418–22.

Drescher, J., N. Blüthgen and H. Feldhaar, 'Population Structure and Intraspecific Aggression in the Invasive Ant Species Anoplolepis gracilis in Malaysian Borneo', *Molecular Ecology*, vol. 16, 2007, pp. 1453–65.

Duckworth, J.W., R.E. Salter and K. Khounboline, *Wildlife in Laos PDR*, 1999 Status Report, Vientiane: IUCN, 1999.

Dunn, R.R., 'Modern Insect Extinctions, the Neglected Majority', *Conservation Biology*, vol. 19, no. 4, 2005, pp. 1030–6.

Durrell, G., *Golden bats and Pink Pigeons: A Journey to the Flora and Fauna of a Unique Island*, New York: Simon and Schuster, 1977.

Dutta, S. et al., *Guidelines for Preparation of State Action Plan for Resident Bustards*, New Delhi: Ministry of Environment and Forests, Government of India, 2013.

Dutta, S., A.R. Rahmani, and Y.V. Jhala, 'Running out of Time? The Great Indian Bustard Ardeotis Nigriceps—Status, Viability, and Conservation Strategies', *European Journal of Wildlife Research*, vol. 57, 2010, pp. 615–25.

Eldredge, Niles, 'The Sixth Extinction', *Action Bioscience* (website), 2001, see http://endangeredink.com/programs/population_and_sustainability/extinction/pdfs/Eldridge-6th-extinction.pdf, accessed 20 December 2022.

Elewa, Ashraf, 'The History, Origins, and Causes of Mass Extinctions', *Journal of Cosmology*, 2009, pp. 201–20.

Elliot, A., J. Del-Hoyo and D. Christie, *Handbook of the Birds of the World*, vol 1: Ostrich to Ducks, Barcelona, Spain: Lynx Edicions, 1992.

Elliott, W., 'Notes on the Great Indian Bustard with Special Reference to the Gular Pocuh', *Proceedings of the Zoological Society of London*, 1880, pp. 486–9.

Ellis, S. et al., eds., *Indian Rhino Vision 2020 Population Modeling Workshop Final Report*, Final Report from the workshop held on 4–5 November 2015, Guwahati, Assam, India, Fort Worth, TX: International Rhino Foundation, 2015.

Ellis, S., 'Enhancing the Survival of the Javan Rhino', *Endangered Species Bulletin*, Spring 2010, pp. 40–1.

Emslie, R. et al., *African and Asian Rhinoceroses: Status, Conservation and Trade*, A Report from the IUCN Species Survival Commission (IUCN/SSC) African and Asian Rhino Specialist Groups and TRAFFIC to the CITES Secretariat pursuant to Resolution Conf. 9.14 (Rev. CoP15). CoP17 Doc. 68 Annex 5, Geneva: CITES Secretariat, 2016.

Emslie, R.H. et al., 'African and Asian Rhinoceroses: Status, Conservation and Trade', Report from the IUCN Species Survival Commission (IUCN SSC) African and Asian Rhino Specialist Groups and Traffic to the CITES Secretariat pursuant to Resolution Conf.9.14 (Rev.CoP17), Columbo, June 2019, 2019, pp. 1–38.

Epstein, H., *Domestic Animals of China*, New York: Africana Publishing Corporation, 1971.

Ernst, C.H. and J.E. Lovich, *Turtles of the United States and Canada*, 2nd edn., Baltimore: Johns Hopkins University Press, 2009.

Evans, W.H., *The Identification of Indian Butterflies*, 2nd edn., Bombay: Bombay Natural History Society, 1932.

Fazl, Abu'l, *Ain-i-Akbari*, tr. Henry Blochmann (1st vol.) and Henry Sullivan Jarrett (2nd and 3rd vols), Calcutta: Asiatic Society of Bengal, 1873.

Feng, L.M. et al., 'First Image of an Amur Leopard Recorded in China', *Cat News*, vol. 55, no. 9, 2011.

Feng, X. et al., 'How Deregulation, Drought and Increasing Fire Impact Amazonian Biodiversity', *Nature*, vol. 597, 2021, pp. 516–21.

Foose, T.J. and N. van Strien, eds., *Asian Rhinos: Status Survey and Conservation Action Plan*, Gland, Switzerland: IUCN/SSC Asian Rhino Specialist Group, 1997.

Franck, P.F., 'Verslag over de jaren 1933–34', *Nederl.-Ind. Vereen. tot Natuurbescherming*, 1934, pp. 40–55.

Galicia, M.P. et al., 'Dietary Habits of Polar Bears in Foxe Basin, Canada: Possible Evidence of a Trophic Regime Shift Mediated by a New Top Predator', *Ecol. Evol.*, vol. 6, 2016, pp. 6005–18.

Galligan, T.H., et al., 'Metabolism of Aceclofenac in Cattle to Vulture-killing Diclofenac', *Conservation Biology: The Journal of the Society for Conservation Biology*, vol. 30, no. 5, 2016, pp. 1122–7.

García, J. et al, 'Spatial Analysis of Habitat Quality in a Fragmented Population of Little Bustard (Tetrax tetrax): Implications for Conservation', *Biological Conservation*, vol. 137, 2007, pp. 45–56.

Gelineau, Kristen, 'Study: Cats, foxes behind Australia's alarming extinctions', *Phys Org* (website), 9 February 2015, see https://phys.org/news/2015-02-cats-foxes-australia-alarming-extinctions.html, accessed 19 January 2023.

'GIB Campaign update: Rajasthan launches "Project Great Indian Bustard"', Conservation India (website), 17 June 2013, see http://www.conservationindia.org/campaigns/gib, accessed 20 December 2022.

Gilbert M. et al., 'Estimating the Potential Impact of Canine Distemper Virus on the Amur Tiger Population (Panthera tigris altaica) in Russia', *PLoS ONE*, vol. 9, no. 10, 2014.

Glatston, A. et al., 'Ailurus fulgens (errata version published in 2017)', The IUCN Red List of Threatened Species (website), 2015, see https://dx.doi.org/10.2305/IUCN.UK.2015-4.RLTS.T714A45195924.en, accessed 4 February 2022.

Gokkon, B., 'Newly Spotted Calves Boost Javan Rhino Population to 72', *Mongabay* (website), 2019, see https://news.mongabay.com/2019/12/javan-rhino-calves-population-72-ujung-kulon-indonesia/, accessed 20 December 2022.

Goossens, B. et al., 'Patterns of Genetic Diversity and Migration in Increasingly Fragmented and Declining Orang-utan (Pongo pygmaeus) Populations from Sabah, Malaysia', Molecular Ecology, vol. 14, no. 2, 2005, pp. 441–56.

Goriup, P.D. and H. Vardhan, *International Symposium on Bustards*, Jaipur: Tourism and Wildlife Society of India, 1980.

Goswami, Dev P., 'Bonoria moh rakkhar guri katha (Katha prasanga)', *Bahi*, vol. 26, no. 10, 1940, pp. 26–9.

Goswami, V.R., V. Vasudev, M.K. and Oli, 'The Importance of Conflict-induced Mortality for Conservation Planning in Areas of Human–elephant Co-occurrence', *Biological Conservation*, vol. 176, 2014, pp. 191–8.

Gray, J.E., *Illustrations of Indian Zoology*, London: Treuttel, Wurtz, and Richter, 1831.

Gray, Louise, 'Polar Bears are Having Fewer Cubs Due to Global Warming', *Telegraph*, 8 February 2011, see http://www.telegraph.co.uk/news/earth/environment/climatechange/8311137/Polar-bears-having-fewer-cubs-due-to-global-warming.html, accessed 19 December 2022.

Gray, T.N. et al, 'Habitat Preferences of a Globally Threatened Bustard Provide Support for Community-based Conservation in Cambodia', *Biological Conservation*, vol. 138, 2007, pp. 341–50.

Greer, D. et al, 'Great Apes Status Report (August 2018)', Report to the CITES Standing Committee on the Status of Great Apes, United Nations Environment Programme Great Apes Survival Partnership, Nairobi, and International Union for Conservation of Nature, Gland, 2018.

Greig, Rebecca, ed., *World of Animals: Book of Big Cats and African Wildlife*, Dorset: Future Publishing, 2018, p. 58.

Greshko, M. and National Geographic Staff, 'What are Mass Extinctions, and What Causes Them?', *National Geographic Science*, 2019, see https://www.nationalgeographic.com/science/article/mass-extinction, accessed 20 December 2022.

Greshko, M., 'What Actually Killed the Dinosaurs? Volcanic Clues Heat up Debate', *National Geographic Science*, 2019, see https://www.nationalgeographic.com/science/article/what-actually-killed-dinosaurs-volcanoes-heat-up-debate, accessed 20 December 2022.

Grimmett, R., C. Inskipp. and T. Inskipp, *Birds of the Indian Subcontinent*, Helm Field Guides, London: A&C Black, 2011.

Groombridge B., *The IUCN Amphibia-Reptilia Red Data Book*, part 1: Testudines, Crocodylia, Rhynchocephalia, Gland: IUCN, 1982.

Guha, Nabarun, 'Heard but Rarely Seen: Plan to Save Assam's White Winged Wood Duck Takes Flight', *Indian Express*, 10 October 2019, see https://indianexpress.com/article/north-east-india/assam/the-call-of-the-spirit-bird-a-move-to-save-assams-elusive-white-winged-wood-duck-6061285/, accessed 17 December 2022.

Gupta, A., 'Jerdon's Courser: A Precious Rarity', *India's Endangered* (website), 1 July 2011, see http://indiasendangered.com/jerdon%E2%80%99s-courser-a-precious-rarity/, accessed 12 May 2020.

Gupta, Alok, 'Carcasses in the Ganga', *The Wire*, April 2016, see https://thewire.in/environment/carcasses-in-the-ganga, accessed 20 December 2022.

Gupta, J., B. Ramesh and Z. Rahman, 'Preserving the Pangolin', *Thirdpole.net*, 2019.

Habung, T., K. Tasser and T. Bamin, 'First record of White-bellied Heron *Ardea insignis* from Kamlang Tiger Reserve, Lohit district, Arunachal Pradesh, India', *Birding ASIA*, vol. 32, 2019, p. 120.

Hallager, S., 'Drinking Methods in Two Species of Bustards', *Wilson Bulletin*, 1994.

Harari, Yuval Noah, 'Industrial Farming is One of the Worst Crimes in History', *The Guardian*, 2015, see https://www.theguardian.com/books/2015/sep/25/industrial-farming-one-worst-crimes-history-ethical-question, accessed 20 December 2022.

Hariyadi, A.R.S. et al., 'Optimizing the Habitat of the Javan Rhinoceros (*Rhinoceros sondaicus*) in Ujung Kulon National Park by Reducing the Invasive Palm *Arenga obtusifolia*', *Pachyderm*, vol. 52, 2012, pp. 49–54.

Harries, Peter and Paul Knorr, 'What does "Lilliput Effect" Mean?', *Palaeogeography, Palaeoclimatology, Palaeoecology*, vol. 284, 2009, pp. 4–10.

Harrington, L.A., N. Cruze, and D. Macdonald, 'Rise to Fame: Events, Media Activity and Public Interest in Pangolins and Pangolin Trade, 2005–2016', *Nature Conservation*, vol. 30, no. 107, 2018.

Haryono, M. et al., 'Monitoring of the Javan Rhino Population in Ujung Kulon National Park, Java', *Pachyderm*, vol. 56, 2015, pp. 1–5.

Haryono, Muh. et al., eds., *Population and Habitat Viability Assessment for the Javan Rhino*, Apple Valley, MN: IUCN/SSC Conservation Breeding Specialist Group, 2016.

Hassanin, A., J.P. Hugot, and B.J. van Vuuren, 'Comparison of Mitochondrial Genome Sequences of Pangolins (Mammalia, Pholidota)', *Comptes rendus biologies*, vol. 338, no. 4, 2015, pp. 260–5.

Hedges, S., *Asian Wild Cattle and Buffaloes: Status Report and Conservation Action Plan*, 2 vols., Working draft for CAMP/AWCSG meeting, Bangkok: WWF Project, 1995.

Heinrich, Sarah et al., 'Where Did All the Pangolins Go? International CITES Trade in Pangolin Species', *Global Ecology and Conservation*, vol. 8, 2016, pp. 241–53.

Hippargi, R., P. Bolde and A. Bolde, 'Ecological Importance of Footprint and Dust Bath Imprints of Great Indian Bustard (Ardeotis nigriceps)', *Ela Journal of Forestry and Wildlife*, vol. 5, 2016, pp. 199–205.

Hippargi, R.V., P.M. Bolde, and A.M. Bolde, 'Notes on the Feeding and Drinking Behaviour of a Female Great Indian Bustard Ardeotis nigriceps', *Ela Journal of Forestry and Wildlife*, vol. 4, 2015, pp. 96–100.

Höglund, J. and R.V. Alatalo, *Leks*, Princeton: Princeton University Press, 1995.

Hommel, Patrick W.F.M., 'A phytosociological study of a forest area in the humid tropics (Ujung Kulon, West Java, Indonesia)', *Vegetatio*, vol. 89, 1990, pp. 39–54.

Hommel, Patrick W.F.M., *Landscape-ecology of Ujung Kulon (West Java, Indonesia)*, n.p.: Patrick W.F.M. Hommel, 1987.

Honacki, J.H., K.E. Kinman and J.W. Koeppl, *Mammal Species of the World: A Taxonomic and Geographic Reference*, Lawrence: Allen Press Inc., and the Association of Systematic Collections, 1982.

Horn, B.K.P., 'Dodo Apocrypha', *Science News*, vol. 113, 1978, p. 19. https://www.wwf.org.uk/learn/wildlife/amur-leopards#:~:text=Adopt%20an%20Amur%20leopard%20today,be%20left%20in%20the%20wild, accessed 5 January 2023.

Hu, Y. et al., 'Genomic Evidence for Two Phylogenetic Species and Long-term Population Bottlenecks in Red Pandas', *Science Advances*, vol. 6, no. 9, 2020.

Hulme, Julian, 'The Dodo: From Extinction to the Fossil Record', *Geology Today*, vol. 28, no. 4, 2012, p. 148.

Hume, A.O. and C.H.T. Marshal, *The Game Birds of India, Burmah and Ceylon*, vol. I, 1878; repr., New Delhi: Bhavna Books and Prints.

Inverarity, J.D., 'The Indian Wild Buffalo', *Journal of the Bombay Natural History Society*, vol. 10, 1895, pp. 41–5.

Inverarity, J.D., 'The Indian Wild Buffalo', *Journal of the Bombay Natural History Society*, vol. 10, 1895, pp. 41–5.

IPBES, *Global Assessment Report on Biodiversity and Ecosystem Services of the Intergovernmental Science-Policy Platform on Biodiversity and Ecosystem Services*, ed. E.S. Brondizio et al., Bonn, Germany: IPBES Secretariat, 2019.

J. del Hoyo, A. Elliott, and J. Sargatal, *Handbook of the Birds of the World: Hoatzin to Auks*, vol. III, Barcelona: Lynx Edicions, 1996.

Jabr, Ferris, 'The Science is In: Elephants are Even Smarter Than We Realized', *Scientific American*, 26 February 2014; Joshua M. Plotnik et al., 'Elephants Know When They Need a Helping Trunk in a Cooperative Task', *Proceedings of the National Academy of Sciences*, vol. 108, no. 12, 2011, pp. 5116–21.

Jackonson, Andrew P. et. al., 'Leopard (*Panthera pardus*) Status, Distribution, and the Research Efforts across its Range', *PeerJ*, vol. 4, 2016.

Jackson, P. and K. Nowell, 'Panthera pardus ssp. orientalis: The IUCN Red List of Threatened Species', ResearchGate, 2016, seehttps://www.researchgate.net/publication/308899520_Panthera_pardus_The_IUCN_Red_List_of_Threatened_Species_2016/citation/download, accessed 25 January 2023.

Jayadevan, Praveen and L. Namassivayan, 'Sightings of Yellow-throated Bulbul Pycnonotus xantholaemus from Chinnar Wildlife Sanctuary, Kerala, Southern India', *Zoos' Print Journal*, vol. 21, no. 4, 2006.

Jeganathan, P., 'Jerdon's Courser', *Nature in Focus: Nature Conservation Foundation* (website), 2017, see https://www.natureinfocus.in/nature-and-wildlife-conservation/jerdon-s-courser, accessed 12 May 2022.

Jeganathan, P. et al., 'Conservation of the Critically Endangered Jerdon's Courser *Rhinoptilus bitorquatus* in India', Mumbai: Bombay Natural History Society, 2004, p. 32.

Jeganathan, P. et al., 'Modelling Habitat Selection and Distribution of the Critically Endangered Jerdon's Courser *Rhinoptilus bitorquatus* in Scrub Jungle: An Application of a New Tracking Method', *Journal of Applied Ecology*, vol. 41, 2002, pp. 224–37.

Jeganathan, P. et al., 'Use of Tracking Strips and Automatic Cameras for Detecting Critically Endangered Jerdon's

Courser *Rhinoptilus bitorquatus* in Scrub Jungle in Andhra Pradesh', *Oryx*, vol. 36, 2002, pp. 182–8.

Jeganathan, P., 'Ecology of the Jerdon's Courser *Rhinoptilus bitorquatus* in India', PhD thesis, University of Mumbai, 2006.

Jerdon, *The Birds of India, being a Natural History of All the Birds Known to Inhabit Continental India; with Descriptions of the Species, Genera, Families, Tribes, and Orders, and a Brief Notice of such Families as are not Found in India, Making it a Manual of Ornithology Specially Adapted for India*, vol. II, part II. Calcutta: T.C. Jerdon, 1864.

Jerdon, Thomas Caverhill, *The Birds of India: Being a Natural History of All the Birds Known to Inhabit Continental India, with Descriptions of the Species, Genera, Families, Tribes, and Orders, and a Brief Notice of Such Families as are Not Found in India, Making it a Manual of Ornithology Specially Adapted for India*, Calcutta: Military Orphan Press, Calcutta, 1863.

Jetz, W. et al., 'Global Distribution and Conservation of Evolutionary Distinctness in Birds', *Current Biology*, vol. 24, 2014, pp. 919–30.

Jiang, Guwangshun et. al., 'New Hope for the Survival of the Amur Leopard in China', *Scientific Report*, vol. 5, 2015.

Johnsgard, P.A., *Bustards, hemipodes, and sandgrouse: birds of dry places*, New York: Oxford University Press, 1991.

Jones, K. et al., 'Global Trends in Emerging Infectious Diseases', *Nature*, vol. 451, 2008, pp. 990–3.

Jordan, K., 'Papilionidae, Papilio to Armandia', *Macrolepidoptera of the World*, vol. 9, 1908–9, pp. 11–109.

Joshi, B.D. et al., 'Siang River in Arunachal Pradesh Splits Red Panda into Two Phylogenetic Species', *Mammalian Biology*, vol. 101, no. 1, 2021, pp. 121–4.

Kanagaraj, R. et al., 'Predicting Range Shifts of Asian Elephants under Global Change', *Diversity and Distributions*, vol. 25, no. 5, 2019, pp. 822–38.

Karthikeyan, S., J.N. Prasad and T.S. Srinivasa, 'Yellow-throated Bulbul Pycnonotus xantholaemus (Jerdon) at Biligirirangan Hills, Karnataka', *J. Bombay Nat. Hist. Soc.*, vol. 92, 1995, pp. 123–4.

Kehimkar, I., *The Book of Indian Butterflies*, Bombay: Bombay Natural History Society and Delhi: Oxford University, 2008.

Khan, A.A. et al., 'Status, threats and conservation of the Great Indian Bustard Ardeotis nigriceps (Vigors) in Pakistan', *Current Science*, 2008, pp. 1079–82 (2008).

Khanal, B., M.K. Chalise and G.S. Solanki, 'Threatened Butterflies of Central Nepal', *Journal of Threatened Taxa*, vol. 5, no. 11, 2013, pp. 4612–15.

King, B.F. et al., 'Birds Recorded during Two Expeditions to North Myanmar (Burma)', *Forktail*, vol. 17, 2001, pp. 29–40.

Kinloch, A.A., *Large Game Shooting in Thibet, the Himalaya, Northern and Central India*, Calcutta: Thacker, Spink & Co., 1885.

———, *Large Game Shooting in Thibet, the Himalaya, Northern and Central India*, 3rd revised and enlarged edn., Calcutta: Thacker, Spink & Co., 1892, p. 307.

Kisor Chaudhuri, 'Impact of Mining on Elephants of Chhotanagpur Plateau, Central India'. *Gajah*, vol. 42, 2015, pp. 30–3.

Knox, A. and S. Piertney, 'The Discovery of the Egg of Jerdon's Courser *Rhinoptilus bitorquatus* (1848) in India', *Journal of the Bombay Natural History Society*, vol. 103, no. 2, 2013, pp. 149–52.

Kokotovic, Marlee, 'The Bee is Officially on the Verge of Extinction', Nation of Change (website), 2019, see https://www.nationofchange.org/2019/12/05/the-bee-is-officially-on-the-verge-of-extinction/, accessed 4 April 2021.

Kormos, T., 'Manis hungrarica, n. sp., des erste Schuppentier aus dem europaischen Oberpliozan', *Folia Zoologica et Hydrobiologica*, vol. 6, 1934, pp. 87–94.

Kotwal, P.C. and R.P. Mishra, 'Ecobiology of Indian Wild Buffalo *Bubalus arnee* L. in Udanti Wildlife Sanctuary, Chhattisgarh, India', *Journal of the Bombay Natural History Society*, vol. 101, no. 2, 2004, pp. 252–4.

Kotwal, P.C., M.L. Naik and R.P. Mishra, 'Indian Wild Buffalo (*Bubalus bubalis* L.) an Endangered Species: Habitat and Diurnal Activity Pattern in Udanti Wildlife Sanctuary', *Cheetal*, vol. 41, nos. 3–4, 2003, pp. 19–28.

Krishna, Nanditha, *Hinduism and Nature*, New Delhi: Penguin Random House, 2017, p. 16.

Kulkarni, Chiranjeevi, '600 elephants electrocuted in India in 10 years', *Deccan Herald*, 17 December 2021, see https://www.deccanherald.com/state/600-elephants-electrocuted-in-india-in-10-years-1061688.html, accessed 15 December 2022.

Kumari, Rima et al., 'Deforestation in India: Consequences and Sustainable Solutions', in *Forest Degradation Around the World*, ed Mohd Nazip Suratman et al., 2019, see https://www.intechopen.com/chapters/66710, accessed 20 December 2022.

Kutz, Cat, 'Building An All-Women Army for Conservation In India: Earth Optimism Speaker Interview with Purnima Devi Barman', *Smithsonian Magazine*, 2020, see https://www.smithsonianmag.com/blogs/earth-optimism/2020/03/06/building-all-woman-army-conservation-india/, accessed 7 April 2020.

Kwapich, Christina,'Leptoptilos dubiusgreater Adjutant', Animal Diversity Web (website), 2011, see https://animaldiversity.org/accounts/Leptoptilos_dubius/#ecosystem_roles, accessed 20 December 2022.

Lackey, Katherine, 'Yellowstones Wolves are Back But They Haven't Restoeed the Parks Ecosystem: here's Why', *USA Today*, 7 September 2018, see https://www.usatoday.com/story/tech/science/2018/09/07/wolves-reintroduction-yellowstone-ecosystem/973658002/, accessed 8 December 2022.

Lahkar, B.P., B.K. Talukdar and P.K. Sarma, 'Invasive Species in Grassland Habitat: An Ecological Threat to Greater One Horned Rhino *Rhinoceros unicornis*', *Pachyderm*, vol. 49, pp. 33–9.

Lents, Nathan H., *Not so Different: Finding Human Nature in Animals*, New York: Columbia University Press, 2016.

Liebing, N. et al., 'Molecular Phylogeny of the Softshell Turtle Genus Nilssonia Revisited, with First Records of N. formosa for China and Wild-living N. Nigricans for Bangladesh', *Vertebrate Zoology*, vol. 62, no. 2, 2012, pp. 261–72

'Livelihoods and Declining Wildlife in the Hkakaborazi National Park, North Myanmar', *Environmental Management*, vol. 48, no. 1, 2011, pp. 158–67.

Livezey B., 'An Ecomorphological Review of the Dodo (Raphus cucullatus) and Solitaire (Pezophaps soliaria), Flightless Columbiformes of the Mascarene Islands', *Journal of Zoology*, vol. 230, 1993, pp. 247–92.

Lovich, E.J. et al., 'Where Have all the Turtles Gone and Why Does it Matter?', *BioScience*, vol. 68, no. 10, 2018.

Lowe, S., M. Browne and S. Boudlejas, '100 of the World's Worst Invasive Alien Species', *Aliens*, vol. 12, 2000, pp. 1–12.

Lunn, N. et al., 'Demography of an Apex Predator at the Edge of its Range: Impacts of Changing Sea Ice on Polar Bears in Hudson Bay', *Ecol. Appl.*, vol. 26, 2016, pp. 1302–20.

Lydekker, R., *The Game Animals of India, Burma, Malaya and Tibet*, *London:* Rowland Ward, 1907.

M. Duarte, Carlos, *Impacts of Global Warming on Polar Ecosystems*, Bilbao, Spain: Fundación BBVA, 2008.

Ma, X. et al., 'Population Genomic Analysis Reveals a Rich Speciation and Demographic History of Orang-utans (Pongo pygmaeus and Pongo abelii)', *PLoS One*, vol. 8, no. 10, 2013.

Mackenzie, M.J.S. and Janet Kear, 'The White-winged Wood Duck', *Wildfowl*, vol. 27, 1976, pp. 5–17.

Mackinnon, J.R. and K. Mackinnon, *Review of the Protected Areas System in the Indo-Malayan Realm*, Gland, Switzerland and Cambridge: IUCN, 1986.

Macneil, Caeleigh, 'How Wolves saved the Foxes, Mice,

and Rivers of Yellowstone National Park', *Earthjustice* (website), see https://earthjustice.org/blog/2015-july/how-wolves-saved-the-foxes-mice-and-rivers-of-yellowstone-national-park, accessed 20 December 2022.

Madhusudan, M.D., 'Living amidst Large Wildlife: Livestock and Crop Depredation by Large Mammals in the Interior Villages of Bhadra Tiger Reserve, South India', *Environmental Management*, vol. 31, no. 4, 2003, pp. 466–75.

Maharajah of Cooch Behar, *Thirty-seven Years of Big Game Shooting in Cooch Behar, the Duars and Assam: A Rough Diary*, Bombay: The Times Press, 1908; repr., *Big Game Shooting in Eastern And North Eastern India*, Delhi: Mittal Publications, 1985.

Maheswaran, G., 'Foraging Ecology of the White-bellied Heron, *Ardea insignis*, in Namdapha Tiger Reserve, Arunachal Pradesh, India', Progress Report to Science and Engineering Research Board, (SERB), Department of Science and Technology (DST), Government of India, 2014.

Maheswaran, Gopinath, 'Walong's White Bellies', *Sanctuary Nature Foundation* (website), see https://sanctuarynaturefoundation.org/article/walong%E2%80%99s-white-bellies, accessed 25 January 2023; first published in *Sanctuary Asia*, vol. 41, no. 8, August 2021.

Mahmood, T., N. Irshad and R. Hussain, 'Habitat Preference and Population Estimates of Indian Pangolin (Manis crassicaudata) in District Chakwal of Potohar Plateau, Pakistan', *Russian Journal of Ecology*, vol. 45, no. 1, 2014, pp. 70–5.

Makayla, 'Is it True That 99.9% of All Species Are Extinct?', 2019, see https://www.encyclopedia.com/daily/is-it-true-that-99-9-of-all-species-are-extinct/, accessed 20 December 2022.

Manakadan, R., 'Ecology of the Great Indian Bustard Ardeotis nigriceps (Vigors) Habitats', MSc thesis, University of Bombay, 1986.

Mathur, P.K., P.K. Malik and P.D. Muley, *Ecology and Population Genetics of Asian Wild Buffalo (Bubalus bubalis) in India*, Project Report, Dehradun: Wildlife Institute of India, 1995.

Mavatur, Anandakumar and Ganesh Raghunathan, 'Fostering Human–elephant Coexistence in the Valparai Landscape, Anamalai Tiger Reserve, Tamil Nadu', in *Human-Wildlife Conflict in the Mountains of SAARC Region Compilation of Successful Management Strategies and Practices*, Thimpu: SAARC Forestry Cente, 2014.

Mayr, E., 'Uncertainty in Science: Is the Giant Panda a Bear or a Raccoon?', *Nature*, vol. 323, 1989, pp. 769–71.

Mazzotti, Frank J., 'The Value of Endangered Species: The Importance of Conserving Biological Diversity', *IFAS Extension*, 2017, see https://edis.ifas.ufl.edu/pdf/UW/UW06400.pdf, accessed 20 December 2022.

McEvoy, J.F. et al., 'Two Sides of the Same Coin: Wildmeat Consumption and Illegal Wildlife Trade at the Crossroads of Asia', *Biological Conservation*, 2019, p. 238.

McKee, Jeffrey Kevin, *Spring Nature: The Conflict between Human Population Growth and Earth's Biodiversity*, New Jersey: Rutgers University Press, 2003.

McKinley, D., 'A History of the Passenger Pigeon in Missouri', *The Auk*, vol. 77, 1960, pp. 399–420.

Miquelle, D.G. et. al., 'Identifying Ecological Corridors for Amur Tigers (Panthera tigris altaica) and Amur Leopards (Panthera pardus orientalis)', *Integr. Zool.*, vol. 10, 2015, pp. 389–402.

Mohapatra, Rajesh Kumar, 'Nearly 6,000 Pangolins in Illegal Wildlife Trade in India since 2009', *WWF India*, 2018, see P.C. Murhy, 'Pangoin fossil', *Arceoploty*, vol. 5, 2001, pp. 23–45.

Mondal, H.S. and G. Maheswaran, 'First Nesting Record of White-bellied Heron (*Ardea insignis*) in Namdapha Tiger Reserve, Arunachal Pradesh, India. Special Report', *Birding ASIA*, vol. 21, 2014, pp. 13–17.

Mukherjee, A.K., *Endangered Animals of India*, Calcutta: Zoological Survey of India, 1982.

Mumby, H.S. and J.M. Plotnik, 'Taking the Elephants' Perspective: Remembering Elephant Behavior, Cognition and Ecology in Human-elephant Conflict Mitigation', *Frontiers in Ecology and Evolution*, vol. 6, 2018, p. 122.

Munjpara, S.B., C.N. Pandey and B. Jethva, 'Habitat Use by the Great Indian Bustard Ardeotis nigriceps (Gruiformes: Otididae) in Breeding and Non-breeding Seasons in Kachchh, Gujarat, India', *Journal of Threatened Taxa*, vol. 5, 2013, pp. 3654–60.

New, T., R. Collins and N. Mark, *Swallowtail Butterflies: An Action Plan for Their Conservation*, IUCN/SSC Species Action Plans Series, Switzerland: IUCN SSC Lepidoptera Specialist Group, 1991, p. 36.

Nijman, V., *An Assessment of Trade in Gibbons and Orang-utans in Sumatra, Indonesia*, Selangor: TRAFFIC Southeast Asia, 2009.

Niyogi, D.G., 'Moved from Temple Ponds to the Wild, black Softshell Turtles Walk towards Revival', *Mongabay* (website), 2020, see https://india.mongabay.com/2020/05/moved-from-temple-ponds-to-the-wild-black-Softshell-Turtles-walk-towards-revival/, accessed 20 December 2022.

Nowak, R.M., *Walker's Mammals of the World*, Baltimore and London: The Johns Hopkins University Press, 1999.

Nuwer, Rachel, 'What Would Happen if All the Worlds' Trees Disappeared', *BBC*, 2019, see https://www.bbc.com/future/article/20190911-what-would-happen-if-all-the-worlds-trees-disappeared, accessed 20 December 2022.

Osipova, L. et al., 'Fencing Solves Human-wildlife Conflict Locally but Shifts Problems Elsewhere: A Case Study using Functional Connectivity Modelling of the African Elephant', *Journal of Applied Ecology*, vol. 55, no. 6, 2018, pp. 2673–84.

Owadally, A.W., 'The Dodo and the Tambalacoque Tree', *Science*, vol. 203, 1979, pp. 1363–4.

Paine, R.T., 'Food Web Complexity and Species Diversity', *American Naturalist*, vol. 100, 1966, pp. 65–75.

Pandey, R.K., 'Habitat Utilization and Diurnal Activity Pattern of Indian Wild Buffalo (*Bubalus bubalis* L.) in Indravati National Park, India: A Study of Habitat/Animal Interaction', *Journal of Tropical Forestry*, vol. 4, no. 3, 1988, pp. 269–80.

Parsons, R.E. and K. Cantlie, 'The Butterflies of the Khasia and Jaintia Hills, Assam', *Journal of the Bombay Natural History Society*, vol. 47, 1948, pp. 498–522.

Peacock, Elizabeth, 'A New Polar Bear Population', *Science*, vol. 376, no. 6599, 2022, pp. 1267–8.

Peeken, I. et al., 'Microplastics in the Marine Realms of the Arctic with Special Emphasis on Sea Ice', *Arctic Report Card*, 2018, see https://www.arctic.noaa.gov/Report-Card, accessed 19 December 2022.

Perkins, Sid, 'Ancient Mass Extinction May Have Shrunk Earth's Creatures', *Science*, 2015, see https://www.science.org/content/article/ancient-mass-extinction-may-have-shrunk-earth-s-creatures, accessed 20 December 2022.

Phukan, Rituraj, 'Biodiversity in a Warming World', in *Climate Abandoned:We're on the Endangered Species List*, ed. J. Cody, Carmel, CA: Writing Endeavors Press, 2019, pp. 11–24.

Pikunov, D.G. et al., 'A Survey of Far Eastern Leopards and Amur Tigers in Southwest Primorye Krai, in 2000', *WCS Russia* (website), 2000, see https://russia.wcs.org/en-us/Wildlife/Amur-Leopards/Ecology.aspx, accessed 14 December 2020.

Piper, Ross, *Extinct Animals: An Encyclopedia of Species That Have Disappeared during Human History*, US: Greenwood Press, 2009, pp. 158–60.

Pitra, C. et al., 'Phylogenetic Relationships and Ancestral Areas of the Bustards (Gruiformes: Otididae), Inferred from Mitochondrial DNA and Nuclear Intron Sequences', Molecular Phylogenetics and Evolution, vol. 23, 2002, pp. 63–74.

Pradhan, R., T. Norbu and P. Frederick, 'Reproduction and Ecology of the world's Rarest Arseid: The White-bellied Heron (*Ardea insignis*) in Bhutan', 31st Annual Meeting of the Water Bird Society, 30 October–3 November 2007.

Pradhan, R., *White-bellied Heron Project*, 2005–2007, unpublished Annual Report, December 2005–December 2006.

Pradhan, S., 'Some Aspects of Ecology of the Red Panda, Ailurus fulgens in the Singhalila National Park, Darjeeling, India', PhD dissertation, North Bengal University, 1999.

Prakash, Vibhu et al., 'Recent Changes in Populations of Critically Endangered Gyps Vultures in India', *Bird Conservation International*, vol. 29, 2017, pp. 1–16.

Prakash, Vibhu et al., 'Recent Changes in Populations of Resident Gyps Vultures in India', *Hist. Soc. Journal of the Bombay Natural History Society*, vol. 104, 2007, pp. 129–35, see https://www.researchgate.net/publication/228495035_Recent_changes_in_populations_of_resident_Gyps_vultures_in_India, accessed 20 December 2022.

Praschag, Peter et al., 'Genetic Evidence for Wild-living Aspideretes Nigricans and a Molecular Phylogeny of South Asian Softshell Turtles (Reptilia: Trionychidae: Aspideretes, Nilssonia)', *Zoologica Scripta*, vol. 36, 2007, pp. 301–10.

Prater, S.H., *Book of Indian Animals*, 2nd edn., Bombay: Bombay Natural History Society, 1948.

Price, M.R. and G.L. Goodman, 'White-bellied Heron (*Ardea insignis*): Conservation Strategy', IUCN Species Survival Commission White-bellied Heron Working Group, part of the IUCN SSC Heron Specialist Group, 2015.

Radhakrishna, B.P. and R. Vaidyanadhan, *Geology of Karnataka*, Bangalore: Geological Society of India, 1994.

Rahmani, A.R., *Threatened Birds of India: Their Conservation Requirements*, Mumbai: Bombay Natural History Society, 2012.

Rahmani, A.R. and R. Manakadan, *Bustard Sanctuaries of India: Strategies for their Conservation and Management*, Mumbai: Bombay Natural History Society, 1988.

Rahmani, A.R., Status of the Great Indian Bustard in Rajasthan-I, Bombay Natural History Society, 1986.

———, 'Flocking Behaviour of a Resident Population of the Great Indian Bustard Ardeotis nigriceps (Vigors)', Revue d'écologie, 1991.

———, 'Need to Start Project Bustards', Bombay Natural History Society, Mumbai, vol. 20, 2006.

———, 'The Godawan Saga: Great Indian Bustards in Decline', *Sanctuary*, vol. 21, 2001, pp. 24–8.

———, 'The Great Indian Bustard', Final Report in the study of ecology of certain endangered species of wildlife and their habitats, Mumbai: Bombay Natural History Society, 1989.

Rajeev, V., 'Man and Nature in Indian Thought', *International Journal for Environmental Rehabilitation and Conservation*, vol. 4, no. 1, 2013, pp. 50–5.

Rangarajan, M. et al., *Gajah: Securing the Future for Elephants in India*, Report of the Elephant Task Force, Delhi: Ministry of Environment and Forests, 2010.

Rao, M. et al., 'Hunting for a Living: Wildlife Trade, Rural Livelihoods and Declining Wildlife in the Hkakaborazi National Park, North Myanmar', *Environmental Management*, vol. 48, no. 1, 2011, pp. 158–67.

Rastogi, Rajat, 'Lantana Invasion Threatens 40 Percent of India's Tiger Habitat, Reports Study, *Mongabay* (website), 2020, see https://india.mongabay.com/2020/08/lantana-invasion-threatens-40-percent-of-indias-tiger-habitat-reports-study/, accessed 20 December 2022.

Reich, Peter et al., 'Effects of Ozone and Acid Rain on White Pine (Pinus strobus) Seedlings Grown in Five Soils. III. Nutrient Relations', *Canadian Journal of Botany*, vol. 66, 2011.

Richardson, W.S.J., 'Correspondence: Wild Buffalo', *Journal of the Bombay Natural History Society*, vol. 4, no. 3, 1889, p. 231.

Ridley, M., 'In Retrospect: The Selfish Gene', *Nature*, vol. 529, no. 7587, 2016, pp. 462–3.

Rijksen, H.D., 'A Field Study on Sumatran Orang Utans

(Pongo pygmaeus abelii, Lesson 1827)', *Ecology, Behaviour and Conservation*, Wageningen: Veenman, 1978

Rinchen Norbu Wangchuk, 'Guilt-Stricken on Killing a Tiger, Padma Shri Officer Went On To Change Their Destiny in India', *The Better India*, 2020, see https://www.thebetterindia.com/223419/padma-shri-kailash-sankhala-project-tiger-kanha-sariska-india-unsung-hero-nor41/, accessed 2 February 2022.

Ritchie, Hannah and Max Roser, 'Biodiversity', *OurWorldInData.org* (website), 2021, see https://ourworldindata.org/biodiversity, accessed 20 December 2022.

Roberts, M.L., 'On the Subspecies of Red Panda Ailurus fulgens', *The Red or Lesser Panda Studbook*, vol. 2, Rotterdam: Stichting Koninklijke Rotterdamse Diergaarde, 1982, pp. 13–24.

Roberts, M.S. and J.L. Gittleman, 'Ailurus fulgens', *Mammalian Species*, vol. 222, 1984.

Rookmaker, L.C., 'The Former Distribution of the Indian Rhinoceros (*Rhinoceros unicornis*) in India and Pakistan', *Journal of the Bombay Natural History Society*, vol. 80, 1984, pp. 555–63.

RSPN, 'The Critically Endangered WBH', ed. R. Pradhan, T. Norbu and P. Frederick, unpublished report, Thimphu: Royal Society for Protection of Nature, 2011.

Safaris, T., 'Claws Out: Things You Dont Know about Claws', *Thomas Safaris* (blog), 2014, see https://thomsonsafaris.com/blog/animal-claws-out/, accessed 20 June 2020.

Sanderson, G.P., *Thirteen years among the wild beasts of India*, 1879; repr., London: W.H. Allen & Co., 1896, p. 387.

Sectionov, Inov, 'Control of Invasive Arenga Palm (*Arenga obtusifolia*) in Habitat Suitable for Javan Rhino (*Rhinoceros sondaicus*) in Ujung Kulon National Park', Presentation to the 2013 International Elephant and Rhino Conservation and Research Symposium in Pittsburgh the USA, 2013.

Sectionov, Inov, 'Rhinos and Global Warming', *Rhino Resources Center* (website), 2018, see http://www.

rhinoresourcecenter.com/pdf_files/120/1202896893.pdf, accessed 20 December 2022.

Sectionov, Inov, 'Translocation rhino in Indonesia', Presentation to the 2007 Translocation Rhino Workshop in Kenya for Producing Guidelines for the In-situ Re-introduction and Translocation of African and Asian Rhinoceros, 2013.

Senapathi, D. et al., 'Use of Remote Sensing to Measure Change in the Extent of Habitat for the Critically Endangered Jerdon's Courser *Rhinoptilus bitorquatus* in India', *Ibis*, vol. 149, no. 2, 2007, pp. 328–37.

Shenfield High School, n.d., see http://www.shenfield.essex.sch.uk/downloads/content/Endangered_Amur_leopards.pdf, accessed 14 December 2020.

Sibley, C.G. and J.E. Ahlquist, *Phylogeny and Classification of Birds: A Study in Molecular Evolution*, New Haven and London: Yale University Press, 1990.

Sileshi, G.W. et al., 'Integrating Ethno-ecological and Scientific Knowledge of Termites for Sustainable Termite Management and Human Welfare in Africa', *Ecology and Society*, vol. 14, no. 1, 2009.

Silverberg, R., *The Auk, the Dodo and the Oryx*, New York: Thomas Y. Crowell, 1967.

Simek, Rudolf, *Dictionary of Northern Mythology*, UK: D.S. Brewer, 1993.

Singh, S., *Impact Report: Disney Conservation Fund*, 2019, unpublished report.

Singh, S., S. Badola and M. Fernandes, *Identification Cards: Tortoises and Freshwater Turtles of India*, India: TRAFFIC, 2019.

Singh, Shelley, 'Some Successful Animal Conservation Projects around the World', *Economic Times*, 2020, see https://economictimes.indiatimes.com/industry/miscellaneous/some-successful-animal-conservation-projects-around-the-world/tiger/slideshow/77312047.cms, accessed 14 January 2021.

Singleton, I. et al., 'Orangutan Population and Habitat

Viability Assessment: Final Report, Apple Valley', MN: IUCN/SSC Conservation Breeding Specialist Group, 2004.

Society, vol. 63, no. 1, 1966, pp. 32–53.

Sodhi, Navjot, Barry Brook and Corey Bradshaw, 'Causes and Consequences of Species Extinctions', in *The Princeton Guide to Ecology*, Berlin: De Gruyter, 2009.

Soewu, D.A. and I.A. Ayodele, 'Utilisation of Pangolin (Manis sps) in Traditional Yorubic Medicine in Ijebu Province, Ogun State, Nigeria', *Journal of Ethnobiology and Ethnomedicine*, vol. 5, no. 1, 2009, p. 39.

Soh, Y.H. et.al., 'Spatial Correlates of Livestock Depredation by Amur Tigers in Hunchun, China: Relevance of Prey Density and Implications for Protected Area Management', *Biological Conservation*, vol. 169, 2014, pp. 117–27.

Soibam, B., 'Recent sightings of Kaiser-I-Hind Teinopalpus imperialis Hope, 1843 (Lepidoptera: Teinopalpani) from Manipur, India', *Journal of Threatened Taxa*, vol. 8, no. 6, 2016, pp. 8930–33.

Sreekar, R. and C. Srinivasulu, 'New Site Record of Yellow-throated Bulbul Pycnonotus xantholaemus from Andhra Pradesh', *Indian Birds*, vol. 5, no. 5, 2010, p. 157.

Srinivasulu, Chelmala, 'Site records of Yellow-throated Bulbul Pycnonotus xantholaemus (Jerdon, 1844) in the Nallamala Hills, Eastern Ghats, Andhra Pradesh, India', *Zoos' Print Journal*, vol. 18, 2003, pp. 1051–2.

Stanford, Craig B. et al., *Turtles and Tortoises are in Trouble*, Current *Biology*, vol. 30, 2020, R721–R735.

State of the Arctic Marine Biodiversity: Key Findings and Advice for Monitoring, Akureyri, Iceland: Conservation of Arctic Flora and Fauna International Secretariat, 2017.

Stork, N.E., 'How Many Species of Insects and Other Terrestrial Arthropods Are There on Earth?', *Annu. Rev. Entomol*, vol. 63, 2018. pp. 31–45.

Subramanya, S. et al., 'A Trip to Thondebhavi in Search of Yellowthroated Bulbul', *Newsletter for Birdwatchers*, vol. 30, no. 11–12, 1990, p. 7.

Subramanya, S., J.N. Prasad and S. Karthikeyan, 'Status, Habitat, Habits and Conservation of Yellow-throated Bulbul *Pycnonotus xantholaemus* (Jerdon) in South India', *J. Bombay Nat. Hist. Soc.*, vol. 103, no. 2–3, May–December 2006.

Sulikhan, Nadezhda S. et al., 'Canine Distemper Virus in a Wild Far Eastern Leopard (Panthera Pardus Orientalis), *Journal of Wildlife Diseases*, vol. 54, no. 1, 2018, pp. 170–4.

Sundaram, Bharath et al., 'Ecology and Impacts of the Invasive Species, Lantana camara, in a Social-Ecological System in South India: Perspectives from Local Knowledge', *Human Ecology*, vol. 40, 2012.

Supriatin, 'Studi kemungkinan adanya pengaruh alelopati langkap (*Arenga obtusifolia* Blumme ex. Mart) terhadap pertumbuhan semai tumbuhan pakan badak Jawa (*Rhinoceros sondaicus* Desmarest, 1822) di Taman Nasional Ujung Kulon', PhD dissertation, Institut Pertanian Bogor, Bogor, Indonesia, 2000.

Swart, J.M., P.R.K. Richardson and J.W.H. Ferguson, 'Ecological Factors affecting the Feeding Behaviour of Pangolins (Manis temminckii)', *Journal of Zoology*, vol. 247, 1999, pp. 281–92.

Swift, Jonathan, *Gulliver's Travels*, 1726; repr., New Delhi: Avenel Books, 1985.

Szulczyk, K.R., 'The Economics of the Malaysian Palm Oil Industry and its Biodiesel Potential', SSRN Electronic Version, 2013, pp. 1–18.

Taggart, M.A. et al., 'Analysis of Nine NSAIDs in Ungulate Tissues Available to Critically Endangered Vultures in India', *Environmental Science & Technology*, vol. 43, no. 12, 2009, pp. 4561–6.

Talbot, G., *The Fauna of British India, including Ceylon and Burma*, vol 1: Butterflies, London: Taylor and Francis, London, 1939.

Talukdar B.K., 'The Current State of Rhino in Assam and Threats in the 21st Century', *Pachyderm*, vol. 29, 2000, pp. 39–47.

Talukdar, B.K. et al, 'Rhinoceros unicornis', *IUCN Red List of Threatened Species*, 2008, pp. 1–17.

Talukdar, B.K. et al., 'Re-introduction of greater one-horned rhino in Manas National Park, Assam, India under Indian Rhino Vision 2020', in *Global Re-introduction Perspectives: Further Case Studies from around the Globe*, ed. P.S. Soorae, Gland, Switzerland: IUCN/SSC Re-introduction Specialist Group, and Abu Dhabi, UAE: Environment Agency Abu Dhabi, 2013, pp 164–7.

Temple, S.A., 'Plant-Animal Mutualism: Coevolution with Dodo Leads to Near Extinction of Plant', *Science*, vol. 197, no. 4306, 1977, pp. 885–6.

Thapa, A. et al., 'Predicting the Potential Distribution of the Endangered Red Panda across its Entire Range using MaxEnt modeling', *Ecology and Evolution*, vol. 8, 2018, pp. 10552–4.

The IUCN Red List of Threatened Species, 2020, see https://www.iucnredlist.org/, accessed 9 June 2020.

The IUCN Red List of Threatened Species, IUCN (website), 2019, see http://www.iucnredlist.org/, accessed 15 December 2017.

Thekaekara, Tarsh, 'Can Elephants and Humans Live Together?', *The Guardian*, 6 March 2017, see https://www.theguardian.com/environment/2017/mar/06/can-elephants-and-humans-live-together, accessed 15 December 2022;

Tigar, B.J. and P.E. Osborne, 'Invertebrate Diet of the Houbara Bustard Chlamydotis [undulata] Macqueenii in Abu Dhabi from Calibrated Faecal Analysis', *Ibis*, vol. 142, 2000, pp. 466–75.

Toomey, Diane, 'From Loathed to Loved: Villagers Rally to Save Greater Adjutant Stork', *Mongabay* (website), 2016, see https://news.mongabay.com/2016/12/from-loathed-to-loved-villagers-rally-to-save-greater-adjutant-stork/, accessed 2 January 2021.

Tordoff, A.W. et al., 'Avifaunal Surveys in the Lowlands of

Kachin State, Myanmar, 2003–2005', *Nat. Hist. Bull. Sim Soc*, vol. 55, no. 2, 2007, pp. 235–306.

TRAFFIC Bulletin, vol. 23, no. 3, 2011, pp. 92–3.

Trageser, S.J. et al, 'Pangolin Distribution and Conservation Status in Bangladesh', *PloS ONE*, vol. 12, no. 4, 2017.

Trishler, H., 'The Anthropocene', N.T.M., vol. 24, 2016, pp. 309–35, see https://link.springer.com/article/10.1007/s00048-016-0146-3#citeas, accessed 20 December 2022.

Tulloch, D.G. and R.T. Litchfield, 'Wallow for Buffalo', *Australian Wild. Res*, vol. 8, 1981, pp. 555–65.

Tulloch, D.G., 'The Feral Swamp Buffaloes of Australia's Northern Territory', in *The Husbandry and Health of the Domestic Buffalo*, ed. R.E. Cockrill, Rome: FAO, 1974, pp. 493–505.

Turtle Conservation Fund, *A Global Action Plan for Conservation of Tortoises and Freshwater Turtles. Strategy and Funding Prospectus*, 2002–2007, Washington, DC: Conservation International and Chelonian Research Foundation, 2002.

University Of California-Davis, 'Invading Ants Disrupt Ecosystem', *ScienceDaily*, 2001, see www.sciencedaily.com/releases/2001/10/011011065354.htm, accessed 11 May 2020.

Varghese, A.O. et al., 'Habitat Suitability Assessment of Ardeotis nigriceps (Vigors) in Great Indian Bustard Sanctuary, Maharashtra (India) Using Remote Sensing and GIS', *Journal of the Indian Society of Remote Sensing*, vol. 44, 2016, pp. 49–57.

Vaughan and P.O. Wiehe, 'Studies on the Vegetation of Mauritius. III. The Structure and Development of the Upland Climax Forest', *Journal of Ecology*, vol. 29, 1941, pp. 127–60.

Vaughan, Adam, 'Ten Conservation Success Stories when Species Came Back from the Brink', *News Scientist*, 2021, see https://www.newscientist.com/article/mg24933223-400-ten-conservation-success-stories-when-species-came-back-from-the-brink/#ixzz7KBttwW6F, accessed 7 February 2022.

Vaughan, R.E., 'Did the Dodo Do It?' Response to Temple', *Animal Kingdom*, 1984.

Venkataswamappa, M. and M.R. Chaitra, 'Observation of Nesting Yellow-throated Bulbuls', *Oriental Bird Club Bulletin*, vol. 30, 1999, pp. 32–3.

Vigors, 'Great Indian Bustard (Ardeotis nigriceps)', *Proceedings of the Zoological Society of London*, 1830–1.

Vince, G., 'Organized Gangs Target Wildlife Trade', New Scientist, 2002.

Vinny, Naidoo et al., 'Toxicity of Non-Steroidal Anti-Inflammatory Drugs to Gyps Vultures: A New Threat from Ketoprofen', *Biology Letters*, vol. 6, no. 3, 2009, pp. 338–42.

Wang, T., et al., 'A Science-based Approach to Guide Amur Leopard Recovery in China', *Biological Conservation*, vol. 210 (B), 2017, pp. 47–55.

Wani, Hamim Mushtaq et al., 'Status of Vultures in India: A Review', *International Journal of Advanced Scientific Research and Management*, vol .3, no. 12, 2018, p. 181.

Weyler, Rex, 'Gaia: Everything on Earth is Connected', *Greenspace.org* (website), 2019, see https://www.greenpeace.org/international/story/24978/gaia-ecology-earth-is-connected-rex-weyler/, accessed 20 December 2022.

Wilson, E., *The Future of Life*, New York: Alfred A. Knopf, 2002, p. 157.

Wohlleben, Peter, *The Secret Network of Nature*, Kindle edn., Random House, 2017.

Wordsworth, William and S.T. Coleridge, *Lyrical Ballads*, London: Penguin, 1786; repr., 2006.

WTI, 'Shared Concerns for Elephants Bring Together African and Asian countries at the E-8 Meet', Wildlife Conservation, 24 May 2011, see retrieved from https://www.conservationindia.org/news/shared-concerns-for-elephants-bring-together-african-and-asian-countries-at-the-e-8-meet, accessed 15 December 2022.

WWF, 'Critically Endangered Amur Leopards Captured on Video', *Phys Org*, 13 July 201, see https://phys.org/news/2011-07-critically-endangered-amur-leopards-captured.html, accessed 19 May 2020.

Xu, Kathy, 'Why Should we Care for wildlife', *Our Better World* (website), 2020, see https://www.ourbetterworld.org/series/a-wild-life/extra-wild/why-we-should-care-wildlife, accessed 27 April 2021.

Yang, Li et al., 'Reconstructing the Historical Distribution of the Amur Leopard (Panthera pardus orientalis) in Northeast China based on Historical Records', *ZooKeys*, vol. 592, 2016, pp. 143–53.

Yonzon, P.B. and M.L. Hunter Jr., 'Conservation of the Red Panda Ailurus Fulgens', *Biological Conservation*, vol. 57, 1991, pp. 1–11.

Zenger, D. Darwin, 'Lyell and Gradualism', *Nature*, vol. 322, 1986, p. 10.

Zhang, F. et al., 'A Note on Captive Breeding and Reproductive Parameters of the Chinese Pangolin, Manis pentadactyla Linnaeus, 1758', *Zoo Keys*, vol. 618, 2016, p. 219.

Zhang, W.E., Y.P. Ryder, and S.J. Brien, 'Molecular Phylogenetics and the Origins of Placental Mammals', *Nature*, vol. 409, no. 614, 2001, p. 6820.

Zimmerman, Michael J. and Ben Bradley, 'Intrinsic vs. Extrinsic Value', *The Stanford Encyclopedia of Philosophy, ed.* Edward N. Zalta, 2019, see https://plato.stanford.edu/archives/spr2019/entries/value-intrinsic-extrinsic/, accessed 20 December 2022.

Editor and Contributors

PRONAMI BHATTACHARYYA is Assistant Professor in English, Royal Global University, Assam. Her areas of specializations are African American literature, travel writing, postcolonial literature, English literary theory, ecofeminism, Anthropocene studies, Indian English writing and theatre studies.

She is the author of two books, one on translation and another on literary theory, as well as several papers in national and international journals of repute. Apart from being an academician, she is a passionate birder and nature enthusiast and has covered more than 450 rare species of birds all over the North-East, Rajasthan, Uttarakhand, West Bengal, and the Nilgiris, some of which are on the verge of extinction. She is currently translating a book from Mizo into English.

AKASH DEEP BARUAH is the Principal Chief Conservator of Forests (PCCF), Tamil Nadu. He specializes in the restoration of degraded mangroves, for which he has developed a successful technique that resulted in 93 per cent increase of mangrove cover between 1999 and 2005. He has published three books on mangrove regeneration, wetland management and Point Calimere Wildlife Sanctuary.

PRERNA SINGH BINDRA is a wildlife conservationist, writer, communications and policy strategist, with a primary focus on conservation of wildlife habitats. She was a member of India's National Board for Wildlife and its core standing committee (2010–13). She has also served on the State Board for Wildlife, Uttarakhand. Prerna is a prolific writer on nature and wildlife in mainstream media, and has authored the critically acclaimed *The Vanishing: India's Wildlife Crisis* as well as *When I Grow up I Want to be a Tiger*, an illustrated book for children. In 2019, she co-edited *Wild Treasures*, an anthology on Natural World Heritage Sites of Asia, jointly published by the Wildlife Institute of India. She is director of Hathi-Sathi Foundation which works to enable safe shared spaces for people and elephants in north-east India.

ANWARUDDIN CHOUDHURY is an Indian naturalist, noted for his expertise on the fauna of north-east India. He is currently the State Coordinator of the Indian Bird Conservation Network. He retired as Divisional Commissioner, Barak valley; Commissioner and Secretary to the Government of Assam, Hill Areas and PPG departments; and Development Commissioner for Hill Areas, Assam. Choudhury rediscovered a rare galliform species, the Manipur Bush Quail in Assam seventy-five years after it was last recorded. He has published twenty-eight books and monographs as well as over 740 articles and scientific papers.

DIPANJOL DEKA is Secretary of the Tea Association of India (TAI). He is a passionate nature enthusiast and birder, having photographed over 700 species all over the north-east, northern and southern part of the country. He is also the Honorary Secretary of the Northeast Writers' Forum (NEWF) apart from being on the board of the Standing Labour Committee, Govt. of Assam; the Regional Advisory Committee of the Central Excise Commissionerate, Shillong; Guwahati Tea Auction Centre; Labour Welfare Board, Govt. of Assam, among others. He is currently authoring a book on wildlife, documenting the (depleting) records of flora and fauna of undivided Darrang district of Assam during the period 1943–66.

RISHIKA DUBLA began her career with the Turtle Survival Alliance/Wildlife Conservation Society's India Turtle Conservation Program in Lucknow. She is currently Development Coordinator with the Wildlife Conservation Society's India Turtle Conservation Program. Rishika's core research interests lie in Darwinian evolutionary biology, herpetology, paleontology and the history of science. She is hopes to bring light to many species' amazing stories of evolution through the medium of art and communication.

T. GANESH is a Senior Fellow at the Suri Sehgal Centre for Biodiversity and Conservation, ATREE. His research interests include faunal loss (birds and small mammals) and its consequence on forest dynamics, long-term changes in tree phenology and factors that contribute to such changes. His current research focuses on lo-term ecological monitoring, and population change in harriers, a migrating grassland raptor, across space and time. He is a part of various advisory panels and is currently a member of the Disaster Management Committee, Tirunelveli district, a member advisor for the Kalakad Mundanthurai Tiger Reserve, and an international advisor for Environmental Studies at New College of Florida, Sarasota, USA.

MONSOON JYOTI GOGOI has been working with the butterflies of north-east India for more than 16 years. After completing his PhD on the butterflies of the Barail Hill range, he is presently part of the Climate Change and Himalayan Program of the Bombay Natural History Society (BNHS).

ULHAS GONDHALI is currently Lecturer at the Jindal Institute of Behavioral Sciences, O.P. Jindal Global University. His work experience ranges from teaching to training forensic professionals. He has previously worked with the Maharashtra

Police, Institute of Forensic Science, Mumbai; Jain University, Bangalore; and Mody University, Jaipur. He has also trained the Mumbai police in crucial forensic techniques and was a key contributor in establishing a 'Crime Scene Simulator' at Maharashtra Police Training Centre, Mumbai. He is a strong advocate of nature conservation and is associated with the Forest Department on various activities concerning wildlife conservation.

S.K. GUPTA is currently Professor and Nodal Officer at the Wildlife Forensic and Conservation Genetics Cell, Wildlife Institute of India, Chandrabani, Dehra Dun. He specializes in forensics, conservation and evolutionary genetics. His research group is currently working on basic and applied research projects to conserve rare and endangered species and strengthen wildlife forensics capability in India. He also teaches on the subjects of wildlife forensics, illegal wildlife trade, conservation and evolutionary genetics, and supervises budding conservation geneticists and forensic researchers. He has professional membership of the IUCN and the International Society of Forensic Genetics.

CAROL INSKIPP is a renowned birder from England. Together with Tim Inskipp and Richard Grimmett, she has co-authored a number of books on birds in South Asia, including several field guides on *Birds of the Indian Subcontinent* (2000; repr., 2011); *Birds of Bhutan* (1999); *Birds of Nepal* (2000; repr., 2016); *Birds of Sri Lanka* (2012); and *Birds of the Eastern Himalayas* (2019). She has also written *A Guide to the Birds of Nepal* (1985, repr., 1991) with her husband Tim and a number of books on bird conservation in Nepal with Hem Sagar Baral including *The State of Nepal's Birds* (2004; repr., 2010); *Important Bird Areas in Nepal* (2005); the *Nepal Bird Red List* (2016); and *Nepal's Forest Birds: Their Status and Conservation* (2019). Carol is currently working on a new edition to the *Birds of the Indian Subcontinent* field guide with Tim and Richard and *Nepal's Wetland Birds: Their Status and Conservation* with Hem.

SARALA KHALING is the Director, Regional Office, Eastern Himalayas, Ashoka Trust for Research in Ecology and the Environment (ATREE), Gangtok (Sikkim). She has been working in the Himalayas since 1994. She has also designed various programmes—regarding conservation, livelihood restoration, climate change mitigation and adaptation—and helped implement them in the field. She has also worked with over a hundred civil society organizations.

RITURAJ PHUKAN is an environmental writer, adventurer and climate change mentor based in Assam. He is currently the National Coordinator for Biodiversity at the Climate Reality Project, India. He is known as Assam's global climate change ambassador. He has been on expeditions to Antarctica and the Arctic region (thrice) to gain first-hand experience of the impacts of climate change on the fragile ecosystems of these regions. Mr Phukan wrote the biodiversity impacts chapter in the book *Climate Abandoned: We're on the Endangered Species List* launched on Earth Day 2019. He is currently working on an initiative to connect and mobilize civil society action in the countries of the Himalayan region. Mr Phukan is also a Member of the International Union for Conservation of Nature (IUCN).

SUNITA PRADHAN is a Conservation Biologist working in the Darjeeling–Sikkim and Himalayan landscape for over twenty years. She previously worked in Darjeeling Zoo (Padmaja Naidu Himalayan Zoological Park) for twelve years, researching the captive upkeep of the Red Panda, Snow Leopard, Himalayan Newt, and other animals. She then joined ATREE, Gangtok. Her research interests focus on understanding and contributing to different events of human–wildlife interactions that lead to their continued co-existence. She is currently researching small, medium, and large mammals along with bats in the Darjeeling–Sikkim region. Her work on tourism in the Red Panda landscape implements appropriate interventions to promote sustainable tourism for Red Panda conservation.

M.B. PRASHANTH works primarily on the study of birds as focal taxa. This includes using self-designed sound playback techniques and understanding the migration patterns of harriers, a migrant grassland raptor in India. He has co-ordinated water bird surveys conducted by ATREE over five years in the Tamiraparni River basin (2010–14). He is also involved in outreach activities include campaigning for a safe and environment friendly annual pilgrimage in Ranthambore National Park and Kalakad Mundunthurai Tiger Reserve.

YOGENDRA SINGH RATHORE works on forest and wildlife conservation with the Rajasthan Forest Department, Government of Rajasthan. He has published several research articles in reputed peer-reviewed international scientific journals. He was awarded the prestigious Fulbright-Nehru Doctoral and Professional Fellowship 2013–14 by the United States–India Education Foundation (USIEF). Later, he was also granted Graduate/Postdoctoral Travel Award 2014 (San Diego, California) by the American Society for Biochemistry and Molecular Biology (ASBMB).

INOV SECTIONOV has been working in the field of rhino conservation in Indonesia for over twenty years. He is Program Coordinator of the International Rhino Foundation (IRF). He served as Program Officer, Indonesia, of Rhino Friend from 2000 to 2006. Inov is involved in several studies and research programs on the Sumatran Rhino in Way Kambas and Bukit Barisan Selatan in collaboration with the Rhino Protection Unit (RPU) and Sumatran Rhino Sanctuary (SRS) while assisting the Rhino Foundation of Indonesia (YABI).

SHAILENDRA SINGH is currently Program Director at Turtle Survival Alliance and Associate Director for Wildlife Conservation Society, India. He is also serving as Regional Vice Chair for South Asia of the IUCN Tortoise and Turtle Specialist Group. Dr Singh has completed over 200 projects for the recovery and research of turtles, crocodiles and River

Dolphins. He was awarded the Conservation Hero Award in 2008 and Lok Samman in 2017 for his innovative approach to conservation.

BIBHAB KUMAR TALUKDAR is an academic and has been working in the field of wildlife research and conservation since 1989. He was one of the founder members and the Chief Executive Officer of Aaranyak, a premier NGO of north-east India. He has been the Chair of Asian Rhino Specialist Group with the Species Survival Commission of International Union for Conservation of Nature (IUCN) since 2008. He has also been the Asia Coordinator of the International Rhino Foundation since June 2008. He has served as a member of the standing committee of the National Board of Wildlife in 2008–10 and the National Board of Wildlife in 2008–17. He is a member of the State Board of Wildlife, Assam and member of the Governing Council of Assam Science Technology and Environment Council. He has published over 60 scientific papers in journals of national and international repute.

Index

www.ingramcontent.com/pod-product-compliance
Lightning Source LLC
Chambersburg PA
CBHW021945170726
48282CB00047B/92/J
9789356878662